Spatiotemporal Modeling of Influenza

Partial Differential Equation Analysis in R

Synthesis Lectures on Biomedical Engineering

Editor
John D. Enderle, *University of Connecticut*

Lectures in Biomedical Engineering will be comprised of 75- to 150-page publications on advanced and state-of-the-art topics that span the field of biomedical engineering, from the atom and molecule to large diagnostic equipment. Each lecture covers, for that topic, the fundamental principles in a unified manner, develops underlying concepts needed for sequential material, and progresses to more advanced topics. Computer software and multimedia, when appropriate and available, are included for simulation, computation, visualization and design. The authors selected to write the lectures are leading experts on the subject who have extensive background in theory, application and design.

The series is designed to meet the demands of the 21st century technology and the rapid advancements in the all-encompassing field of biomedical engineering that includes biochemical processes, biomaterials, biomechanics, bioinstrumentation, physiological modeling, biosignal processing, bioinformatics, biocomplexity, medical and molecular imaging, rehabilitation engineering, biomimetic nano-electrokinetics, biosensors, biotechnology, clinical engineering, biomedical devices, drug discovery and delivery systems, tissue engineering, proteomics, functional genomics, and molecular and cellular engineering.

Spatiotemporal Modeling of Influenza: Partial Differential Equation Analysis in R
William E. Schiesser
2019

PDE Models for Atherosclerosis Computer Implementation in R
William E. Schiesser
2018

Computerized Analysis of Mammographic Images for Detection and Characterization Breast Cancer
Paola Casti, Arianna Mencattini, Marcello Salmeri, and Rangaraj M. Rangayyan
2017

Models of Horizontal Eye Movements: Part 4, A Multiscale Neuron and Muscle Fiber-Based Linear Saccade Model
Alireza Ghahari and John D. Enderle
2015

Bioinstrumentation
John D. Enderle
2006

Fundamentals of Respiratory Sounds and Analysis
Zahra Moussavi
2006

Advanced Probability Theory for Biomedical Engineers
John D. Enderle, David C. Farden, and Daniel J. Krause
2006

Intermediate Probability Theory for Biomedical Engineers
John D. Enderle, David C. Farden, and Daniel J. Krause
2006

Basic Probability Theory for Biomedical Engineers
John D. Enderle, David C. Farden, and Daniel J. Krause
2006

Sensory Organ Replacement and Repair
Gerald E. Miller
2006

Artificial Organs
Gerald E. Miller
2006

Signal Processing of Random Physiological Signals
Charles S. Lessard
2006

Image and Signal Processing for Networked E-Health Applications
Ilias G. Maglogiannis, Kostas Karpouzis, and Manolis Wallace
2006

Spatiotemporal Modeling of Influenza: Partial Differential Equation Analysis in R

William E. Schiesser

ISBN: 978-3-031-00536-7   paperback
ISBN: 978-3-031-01664-6   ebook
ISBN: 978-3-031-00044-7   hardcover

DOI 10.1007/978-3-031-01664-6

A Publication in the Springer Nature series
SYNTHESIS LECTURES ON ADVANCES IN AUTOMOTIVE TECHNOLOGY

Lecture #57
Series Editor: John D. Enderle, *University of Connecticut*
Series ISSN
Print 1930-0328   Electronic 1930-0336

Spatiotemporal Modeling of Influenza

Partial Differential Equation Analysis in R

William E. Schiesser
Lehigh University

SYNTHESIS LECTURES ON BIOMEDICAL ENGINEERING #57

ABSTRACT

This book has a two-fold purpose:

(1) An introduction to the computer-based modeling of influenza, a continuing major world-wide communicable disease.

(2) The use of (1) as an illustration of a methodology for the computer-based modeling of communicable diseases.

For the purposes of (1) and (2), a basic influenza model is formulated as a system of partial differential equations (PDEs) that define the spatiotemporal evolution of four populations: susceptibles, untreated and treated infecteds, and recovereds. The requirements of a well-posed PDE model are considered, including the initial and boundary conditions. The terms of the PDEs are explained.

The computer implementation of the model is illustrated with a detailed line-by-line explanation of a system of routines in R (a quality, open-source scientific computing system that is readily available from the Internet). The R routines demonstrate the straightforward numerical solution of a system of nonlinear PDEs by the method of lines (MOL), an established general algorithm for PDEs.

The presentation of the PDE modeling methodology is introductory with a minumum of formal mathematics (no theorems and proofs), and with emphasis on example applications. The intent of the book is to assist in the initial understanding and use of PDE mathematical modeling of communicable diseases, and the explanation and interpretation of the computed model solutions, as illustrated with the influenza model.

KEYWORDS

communicable disease, influenza, computer-based mathematical model, partial differential equation (PDE), method of lines (MOL), R coding, spatiotemporal solutions, traveling wave solutions

Contents

Preface

This book has a two-fold purpose:

(1) An introduction to the computer-based modeling of influenza, a continuing major world-wide communicable disease.

(2) The use of (1) as an illustration of a methodology for the computer-based modeling of communicable diseases.

For the purposes of (1) and (2), the influenza model by Zhang [2] was selected which has the following general features:

- The model is an introduction to the quantitative analysis of influenza.

- It has the basic elements of a model for communicable diseases such as transmission between susceptibles and untreated/treated infecteds.

- Through the partial differential equation (PDE) formulation it provides the spatiotemporal features of disease evolution.

- Since it is basic (with minimal complexity), the model is well suited as an introduction to numerical methods and computer implementation of PDE models.

- The methodology for PDE model formulation and computer implementation is illustrated with a detailed derivation of the influenza model equations and a line-by-line explanation of associated routines in R^1.

- The R routines are available from a download so that the reader/analyst/researcher can readily execute them to confirm the solutions reported in the book, then experiment with the model, for example, by varying parameter values and PDE terms, on modest computers.

In summary, the presentation of the PDE modeling methodology is introductory with a minumum of formal mathematics (no theorems and proofs), and with emphasis on example applications. The intent of the book is to assist in the initial understanding and use of mathematical modeling, and the explanation and interpretation of the computed model solutions, as illustrated with the influenza model.

---

[1]R is a quality, open-source scientific computing system that is readily available from the Internet [1].

I hope these objectives are fulfilled, and I would welcome hearing about experiences with the use of the book (directed to `wes1@lehigh.edu`).

William E. Schiesser
May 2019

REFERENCES

[1] Soetaert, K., Cash, J., and Mazzia, F. (2012). *Solving Differential Equations in R*, Springer-Verlag, Heidelberg, Germany. DOI: 10.1007/978-3-642-28070-2. xiii

[2] Zhang, T., and W. Wang (2014). Existence of traveling wave solutions for influenza model with treatment, *Journal of Mathematical Analysis and Applications*, 419, pp. 469–495 DOI: 10.1016/j.jmaa.2014.04.068. xiii

CHAPTER 1

PDE Model Formulation

INTRODUCTION

A mathematical model is presented in this book for influenza [2] that includes the spatiotemporal features and treatment of this communicable disease. The model consists of four partial differential equations (PDEs) with the dependent variables (Table 1.1).

Table 1.1: PDE dependent variables

Dependent Variable	Population Per Unit Area
S	Susceptibles
I_u	Untreated infecteds
I_h	Treated infecteds
R	Recovereds

The independent variables for these dependent variables are: (1) r, radial position in a region affected by influenza and (2) t, time for the evolution of the influenza.

The derivation of the PDEs is presented in the next section followed by the initial conditions (ICs) and boundary conditions (BCs) in subsequent sections.

1.1 PDE DERIVATION

The PDE for $S(r,t)$ is based on conservation applied to an incremental area[1] $2\pi r dr$,

$$2\pi r dr \frac{\partial S}{\partial t} = 2\pi r q_r|_r - 2\pi(r+dr)q_r|_{r+dr} - 2\pi r dr \beta(I_u + \delta I_h)S. \tag{1.a}$$

The terms in Eq. (1.a) are briefly explained next.

- $2\pi r dr \frac{\partial S}{\partial t}$: accumulation (when $\frac{\partial S}{\partial t}$ is positive) or depletion (when $\frac{\partial S}{\partial t}$ is negative) of the susceptibles in the incremental area $2\pi r dr$. As an illustrative selection of units, if the spatial scale is in km (kilometers) and the time scale is in days, this term has the

---

[1]The incremental area is the difference of two areas of radius r and $r+dr$, that is, $\pi(r+dr)^2 - \pi r^2 = \pi(r^2 + 2rdr + dr^2 - r^2) = \pi(2rdr + dr^2) \approx 2\pi r dr$ for small dr.

units (km)(km)(susceptibles/km^2)(1/day) = susceptibles/day which is rate of change of the number (population) of susceptibles in the incermental area $2\pi r dr$. For consistency, the other terms in Eq. (1.a) should have the same units.

- $2\pi r q_r|_r$: flux of susceptibles into the incremental area $2\pi r dr$ at r with the units (km)(susceptibles/km)(1/day) = susceptibles/day.

- $-2\pi(r + dr)q_r|_{r+dr}$: flux of susceptibles out of the incremental area $2\pi r dr$ at $r + dr$ with the units (km)(susceptibles/km)(1/day) = susceptibles/day.

- $-2\pi r dr \beta(I_u + \delta I_h)S$: rate of change of susceptibles, $S(r,t)$, resulting from the untreated infected, $I_u(r,t)$ and treated infected, $I_h(r,t)$ in the incremental area $2\pi r dr$. β is the transmission rate of susceptibles to infected. δ is the reduction factor in infectiousness due to antivirial treatment. $\beta(I_u + \delta I_h)$ has the units (1/day).

Division of Eq. (1.a) by $2\pi r dr$ and minor rearrangement gives

$$\frac{\partial S}{\partial t} = -\frac{1}{r}\left(\frac{(r + dr)q_r|_{r+dr} - r q_r|_r}{dr}\right) - \beta(I_u + \delta I_h)S. \tag{1.b}$$

With $dr \to 0$, Eq. (1.b) becomes a PDE in r and t.

$$\frac{\partial S}{\partial t} = -\frac{1}{r}\frac{\partial(r q_r)}{\partial r} - \beta(I_u + \delta I_h)S. \tag{1.c}$$

The flux q_r is given by Fick's first law

$$q_r = -d_s\frac{\partial S}{\partial r}, \tag{1.d}$$

where d_s is a diffusvity with units km^2/day (a constant). Substitution of Eq. (1.d) in Eq. (1.c) gives

$$\frac{1}{r}\frac{\partial}{\partial r}\left(r d_s\frac{\partial S}{\partial r}\right) = d_s\left(\frac{\partial^2 S}{\partial r^2} + \frac{1}{r}\frac{\partial S}{\partial r}\right). \tag{1.e}$$

If the radial group in Eq. (1.e) is expanded, the first of four PDEs that defines $S(r,t)$ results.

$$\frac{\partial S}{\partial t} = d_s\left(\frac{\partial^2 S}{\partial r^2} + \frac{1}{r}\frac{\partial S}{\partial r}\right) - \beta(I_u + \delta I_h)S. \tag{1.1-1}$$

Since the RHS of Eq. (1.1-1) includes I_u and I_h, additional PDEs are required

Conservation applied to untreated infected with population $I_u(r,t)$ over the incremental area $2\pi r dr$ gives the second PDE of the model.

$$\frac{\partial I_u}{\partial t} = d_u\left(\frac{\partial^2 I_u}{\partial r^2} + \frac{1}{r}\frac{\partial I_u}{\partial r}\right) + (1 - \mu)\beta(I_u + \delta I_h)S - k_u I_u, \tag{1.1-2}$$

d_u is the radial diffusivity for untreated infected with units km^2/day, μ is the fraction of new infected cases who are treated, and k_u is the recovery rate of untreated infected with units (1/day).

Conservation applied to treated infected with population $I_h(r, t)$ over the incremental area $2\pi r dr$ gives the third PDE of the model.

$$\frac{\partial I_h}{\partial t} = d_h \left(\frac{\partial^2 I_h}{\partial r^2} + \frac{1}{r} \frac{\partial I_h}{\partial r} \right) + \mu \beta (I_u + \delta I_h) S - k_h I_h, \qquad (1.1\text{-}3)$$

d_h is the radial diffusivity for treated infected with units km^2/day, and k_h is the recovery rate of treated infected with units (1/day).

Conservation applied to recovereds with population $R(r, t)$ over the incremental area $2\pi r dr$ gives the fourth PDE of the model.

$$\frac{\partial R}{\partial t} = d_r \left(\frac{\partial^2 R}{\partial r^2} + \frac{1}{r} \frac{\partial R}{\partial r} \right) + k_u I_u + k_h I_h, \qquad (1.1\text{-}4)$$

d_R is the radial diffusivity for recovered with units km^2/day.

Equations (1.1-1)–(1.1-4) constitute the PDE model for the dependent variables of Table 1.1. The ICs are considered next.

1.2   INITIAL CONDITIONS

Equations (1.1-1)–(1.1-4) are first order in t and therefore each requires one IC.

$$S(r, t = 0) = S_0(r) \qquad (1.2\text{-}1)$$
$$I_u(r, t = 0) = I_{u0}(r) \qquad (1.2\text{-}2)$$
$$I_h(r, t = 0) = I_{h0}(r) \qquad (1.2\text{-}3)$$
$$R(r, t = 0) = R_0(r), \qquad (1.2\text{-}4)$$

where $S_0(r), I_{u0}(r), I_{h0}(r), R_0(r)$ are functions to be specified.

1.3   BOUNDARY CONDITIONS

Equations (1.1-1)–(1.1-4) are second order in r and therefore each requires two BCs. As the first case, homogeneous (zero) Neumann BCs are specified.

$$\frac{\partial S(r = r_l, t)}{\partial r} = \tag{1.3-1}$$

$$\frac{\partial S(r = r_u, t)}{\partial r} = 0 \tag{1.3-2}$$

$$\frac{\partial I_u(r = r_l, t)}{\partial r} = \tag{1.3-3}$$

$$\frac{\partial I_u(r = r_u, t)}{\partial r} = 0 \tag{1.3-4}$$

$$\frac{\partial I_h(r = r_l, t)}{\partial r} = \tag{1.3-5}$$

$$\frac{\partial I_h(r = r_u, t)}{\partial r} = 0 \tag{1.3-6}$$

$$\frac{\partial R(r = r_l, t)}{\partial r} = \tag{1.3-7}$$

$$\frac{\partial R(r = r_u, t)}{\partial r} = 0, \tag{1.3-8}$$

r_l, r_u are lower and upper boundary values of r to be specified, that is, $r_l \leq r \leq r_u$.

Equations (1.1-1)–(1.1-4), (1.2-1)–(1.2-4), and (1.3-1)–(1.3-8) constitute the influenza model. The implementation of this model with programming in R^2 considered in Chapter 2.

1.4   SUMMARY AND CONCLUSIONS

An influenza model expressed as a system of four PDEs is derived in this chapter. This model is then integrated (solved) by the MOL implemented as a series of routines in R. Experimentation with the model, such as variation of (1) the parameters and (2) the ICs and BCs is illustrated in subsequent chapters.

REFERENCES

[1] Soetaert, K., Cash, J., and Mazzia F. (2012). *Solving Differential Equations in R*, Springer-Verlag, Heidelberg, Germany. DOI: 10.1007/978-3-642-28070-2. 4

[2] Zhang, T. and Wang, W. (2014). Existence of traveling wave solutions for influenza model with treatment, *Journal of Mathematical Analysis and Applications*, 419, pp. 469–495. DOI: 10.1016/j.jmaa.2014.04.068. 1

---

[2]R is an open-source, quality scientific programming system that is readily available for download from the Internet. In particular, the R library integrators for ordinary differential equations (ODEs) [1] are used for the solution of the PDE model by the method of lines (MOL) as explained in Chapter 2.

CHAPTER 2

Model Implementation

INTRODUCTION

The computer implementation of the PDE influenza model developed in Chapter 1 is discussed in this chapter. The implementation is formulated as a set of routines that can be compiled and executed within the basic R system.[1] The discussion of the R implementation starts with a main program.

2.1 MAIN PROGRAM

A main program for Eqs. (1.1-1)–(1.1-4), (1.2-1)–(1.2-4), and (1.3-1)–(1.3-8), follows.

Listing 2.1: Main program for Eqs. (1.1-1)–(1.1-4), (1.2-1)–(1.2-4), and (1.3-1)–(1.3-8)

```
#
# Influenza model
#
# Delete previous workspaces
  rm(list=ls(all=TRUE))
#
# Access ODE integrator
  library("deSolve");
#
# Access functions for numerical solution
  setwd("f:/flu/chap2");
  source("pde1a.R");
  source("dss004.R");
  source("dss044.R");
#
# Parameters for ODEs
  ds=1;
  du=1;
  dh=1;
```

[1]R is a quality, open-source scientific computing system that is readily available from the Internet [1]. The Rstudio editor is recommended when using R and producing graphical (plotted) output in the standard formats, e.g., png, pdf.

```
  dr=1;
  beta=0.008;
  delta=0.6
  mu=0.5;
  ku=0.1;
  kh=0.2;
#
# Spatial grid (in r)
  nr=51;
  rl=0;ru=100;
  r=seq(from=rl,to=ru,by=(ru-rl)/(nr-1));
#
# Independent variable for ODE integration
  t0=0;tf=300;nout=16;
  tout=seq(from=t0,to=tf,by=(tf-t0)/(nout-1));
#
# Initial condition (t=0)
  ncase=1;
  u0=rep(0,4*nr);
  if(ncase==1){
    for(i in 1:nr){
      u0[i]      =50;
      u0[i+nr]   =0;
      u0[i+2*nr]=0;
      u0[i+3*nr]=0;
    }
  }
  if(ncase==2){
    for(i in 1:nr){
      u0[i]      =50;
      u0[i+nr]   =5*exp(-0.01*r[i]^2);
      u0[i+2*nr]=0;
      u0[i+3*nr]=0;
    }
  }
  ncall=0;
#
# ODE integration
  out=lsodes(y=u0,times=tout,func=pde1a,
```

```
      sparsetype="sparseint",rtol=1e-6,
      atol=1e-6,maxord=5);
  nrow(out)
  ncol(out)
#
# Arrays for plotting numerical solution
   S=matrix(0,nrow=nr,ncol=nout);
  Iu=matrix(0,nrow=nr,ncol=nout);
  Ih=matrix(0,nrow=nr,ncol=nout);
   R=matrix(0,nrow=nr,ncol=nout);
  for(it in 1:nout){
    for(i in 1:nr){
       S[i,it]=out[it,i+1];
      Iu[i,it]=out[it,i+1+nr];
      Ih[i,it]=out[it,i+1+2*nr];
       R[i,it]=out[it,i+1+3*nr];
    }
  }
#
# Display numerical solution
  for(it in 1:nout){
    if((it-1)*(it-nout)==0){
      cat(sprintf("\n"));
      cat(sprintf("\n        t           r"));
      cat(sprintf("\n                   S(r,t)        Iu(r,t)"));
      cat(sprintf("\n                   Ih(r,t)         R(r,t)"));
    for(i in 1:nr){
      if((i-1)*(i-26)*(i-nr)==0){
      cat(sprintf("\n %6.0f %6.1f",tout[it],r[i]));
      cat(sprintf("\n          %12.2f %12.2f", S[i,it],Iu[i,it]));
      cat(sprintf("\n          %12.2f %12.2f",Ih[i,it], R[i,it]));
      }
    }
    }
  }
#
# Calls to ODE routine
  cat(sprintf("\n\n ncall = %5d\n\n",ncall));
#
```

```
# Plot PDE solutions
#
# S
  par(mfrow=c(1,1));
  matplot(r,S,type="l",xlab="r",ylab="S(r,t)",
    lty=1,main="",lwd=2,col="black");
#
# Iu
  par(mfrow=c(1,1));
  matplot(r,Iu,type="l",xlab="r",ylab="Iu(r,t)",
    lty=1,main="",lwd=2,col="black");
#
# Ih
  par(mfrow=c(1,1));
  matplot(r,Ih,type="l",xlab="r",ylab="Ih(r,t)",
    lty=1,main="",lwd=2,col="black");
#
# R
  par(mfrow=c(1,1));
  matplot(r,R,type="l",xlab="r",ylab="R(r,t)",
    lty=1,main="",lwd=2,col="black");
```

We can note the following details about Listing 2.1.

- Previous workspaces are deleted.

```
#
# Influenza model
#
# Delete previous workspaces
  rm(list=ls(all=TRUE))
```

- The R ODE integrator library deSolve is accessed. Then the directory with the files for the solution of Eqs. (1.1-1)–(1.1-4), (1.2-1)–(1.2-4), and (1.3-1)–(1.3-8) is designated. Note that setwd (set working directory) uses / rather than the usual \.

```
#
# Access ODE integrator
  library("deSolve");
#
```

```
# Access functions for numerical solution
  setwd("f:/flu/chap2");
  source("pde1a.R");
  source("dss004.R");
  source("dss044.R");
```

pde1a.R is the routine for the method of lines (MOL) approximation of PDEs (1.1) (discussed subsequently). dss004, dss044 (Differentiation in Space Subroutine) are library routines for calculating first and second derivatives in r, respectively.

• The model parameters are defined numerically.

```
#
# Parameters for ODEs
  ds=1;
  du=1;
  dh=1;
  dr=1;
  beta=0.008;
  delta=0.6
  mu=0.5;
  ku=0.1;
  kh=0.2;
```

These parameter values are taken from [2] with the exception of the diffusivities, ds, du, dl, dr which were increased to give additional smoothing (diffusion, dispersion) in r. This was required since the lower values in [2] give solutions that are spatially irregular (the reader can verify this conclusion as an exercise).

• A spatial grid of 51 points is defined for $r_l = 0 \le r \le r_u = 100$, so that $r = 0, 100/(51 - 1) = 2, \ldots, 100$.

```
#
# Spatial grid (in r)
  nr=51;
  rl=0;ru=100;
  r=seq(from=rl,to=ru,by=(ru-rl)/(nr-1));
```

The upper boundary $r_u = 100$ can represent a circular (polar) region with radius 100 km, and the PDE dependent variables of Table 1.1 are the number of influenza cases over a 1 km^2 area in this region. nr=51 was selected to give a smooth solution in r while maintaining a manageable number of MOL ODEs.

- An interval in t of 16 points is defined for $0 \le t \le 300$ days so that $tout = 0, 300/(16 - 1) = 20, \ldots, 300$.

```
#
# Independent variable for ODE integration
  t0=0;tf=300;nout=16;
  tout=seq(from=t0,to=tf,by=(tf-t0)/(nout-1));
```

nout=16 was selected to give a clear indication of the movement of the PDE solutions in t as reflected in the graphical output that follows.

- ICs (1.2) are defined for two cases.

```
#
# Initial condition (t=0)
  ncase=1;
  u0=rep(0,4*nr);
  if(ncase==1){
    for(i in 1:nr){
      u0[i]      =50;
      u0[i+nr]   =0;
      u0[i+2*nr]=0;
      u0[i+3*nr]=0;
    }
  }
  if(ncase==2){
    for(i in 1:nr){
      u0[i]      =50;
      u0[i+nr]   =5*exp(-0.01*r[i]^2);
      u0[i+2*nr]=0;
      u0[i+3*nr]=0;
    }
  }
  ncall=0;
```

For ncase=1, the susceptibles start at $S(r, t = 0) = 50$, and the other three populations start with no cases, $I_u(r, t = 0) = I_h(r, t = 0) = R(r, t = 0) = 0$. For ncase=2 a region around $r = 0$ starts with an untreated infected population $I_u(r, t = 0)$ defined by a Gaussian function centered at $r = 0$.

u0 therefore has $(4)(51) = 204$ elements. The counter for the calls to the ODE/MOL routine pde1a is also initialized.

- The system of 204 MOL/ODEs is integrated by the library integrator lsodes (available in deSolve). As expected, the inputs to lsodes are the ODE function, pde1a, the IC vector u0, and the vector of output values of t, tout. The length of u0 (204) informs lsodes how many ODEs are to be integrated. func,y,times are reserved names.

```
#
# ODE integration
  out=lsodes(y=u0,times=tout,func=pde1a,
      sparsetype="sparseint",rtol=1e-6,
      atol=1e-6,maxord=5);
  nrow(out)
  ncol(out)
```

The numerical solution to the ODEs is returned in matrix out. In this case, out has the dimensions $nout \times (4nr + 1) = 16 \times 4(51) + 1 = 204 + 1 = 205$, which are confirmed by the output from nrow(out),ncol(out) (included in the numerical output considered subsequently).

The offset $+ 1$ is required since the first element of each column has the output t (also in tout), and the $2, \ldots, 4nr + 1 = 2, \ldots, 205$ column elements have the 204 ODE solutions.

- The solutions of the 204 ODEs returned in out by lsodes are placed in arrays S,Iu,Ih,R.

```
#
# Arrays for plotting numerical solution
   S=matrix(0,nrow=nr,ncol=nout);
  Iu=matrix(0,nrow=nr,ncol=nout);
  Ih=matrix(0,nrow=nr,ncol=nout);
   R=matrix(0,nrow=nr,ncol=nout);
  for(it in 1:nout){
    for(i in 1:nr){
       S[i,it]=out[it,i+1];
      Iu[i,it]=out[it,i+1+nr];
      Ih[i,it]=out[it,i+1+2*nr];
       R[i,it]=out[it,i+1+3*nr];
    }
  }
```

Again, the offset i+1 is required since the first element of each column of out has the value of t.

- $S(r,t)$, $I_u(r,t)$, $I_h(r,t)$, $R(r,t)$ are displayed numerically as a function of r and t for $t = 0, 300$, $r = 0, 50, 100$.

```
#
# Display numerical solution
  for(it in 1:nout){
    if((it-1)*(it-nout)==0){
      cat(sprintf("\n"));
      cat(sprintf("\n        t          r"));
      cat(sprintf("\n                  S(r,t)        Iu(r,t)"));
      cat(sprintf("\n                  Ih(r,t)        R(r,t)"));
      for(i in 1:nr){
        if((i-1)*(i-26)*(i-nr)==0){
        cat(sprintf("\n %6.0f %6.1f",tout[it],r[i]));
        cat(sprintf("\n            %12.2f %12.2f", S[i,it],Iu[i,it]));
        cat(sprintf("\n            %12.2f %12.2f",Ih[i,it], R[i,it]));
        }
      }
    }
  }
```

- The number of calls to pde1a is displayed at the end of the solution.

```
#
# Calls to ODE routine
  cat(sprintf("\n\n ncall = %5d\n\n",ncall));
```

- $S(r,t)$, $I_u(r,t)$, $I_h(r,t)$, $R(r,t)$ are plotted as a function of r with t as a parameter.

```
#
# Plot PDE solutions
#
# S
  par(mfrow=c(1,1));
  matplot(r,S,type="l",xlab="r",ylab="S(r,t)",
    lty=1,main="",lwd=2,col="black");
#
# Iu
  par(mfrow=c(1,1));
  matplot(r,Iu,type="l",xlab="r",ylab="Iu(r,t)",
    lty=1,main="",lwd=2,col="black");
```

```r
#
# Ih
  par(mfrow=c(1,1));
  matplot(r,Ih,type="l",xlab="r",ylab="Ih(r,t)",
    lty=1,main="",lwd=2,col="black");
#
# R
  par(mfrow=c(1,1));
  matplot(r,R,type="l",xlab="r",ylab="R(r,t)",
    lty=1,main="",lwd=2,col="black");
```

This completes the discussion of the main program of Listing 2.1. The MOL/ODE routine pde1a called by lsodes is considered next.

2.2   ODE/MOL ROUTINE

The ODE/MOL routine called by lsodes follows.

Listing 2.2: ODE/MOL routine for Eqs. (1.1-1)–(1.1-4), (1.2-1)–(1.2-4), and (1.3-1)–(1.3-8)

```r
  pde1a=function(t,u,parms){
#
# Function pde1a computes the t derivative
# vectors of S(r,t),Iu(r,t),Ih(r,t),R(r,t)
#
# One vector to four vectors
  S=rep(0,nr);Iu=rep(0,nr);
  Ih=rep(0,nr); R=rep(0,nr);
  for(i in 1:nr){
    S[i]=u[i];
   Iu[i]=u[i+nr];
   Ih[i]=u[i+2*nr];
    R[i]=u[i+3*nr];
  }
#
# Algebra
  rate=rep(0,nr);
  for(i in 1:nr){
    rate[i]=beta*(Iu[i]+delta*Ih[i])*S[i];
  }
#
```

```
# Sr,Iur,Ihr,Rr
   Sr=dss004(rl,ru,nr, S);
  Iur=dss004(rl,ru,nr,Iu);
  Ihr=dss004(rl,ru,nr,Ih);
   Rr=dss004(rl,ru,nr, R);
#
# BCs
   Sr[1]=0; Sr[nr]=0;
  Iur[1]=0;Iur[nr]=0;
  Ihr[1]=0;Ihr[nr]=0;
   Rr[1]=0; Rr[nr]=0;
#
# Srr,Iurr,Ihrr,Rrr
  nl=2;nu=2;
   Srr=dss044(rl,ru,nr, S, Sr,nl=2,nu=2);
  Iurr=dss044(rl,ru,nr,Iu,Iur,nl=2,nu=2);
  Ihrr=dss044(rl,ru,nr,Ih,Ihr,nl=2,nu=2);
   Rrr=dss044(rl,ru,nr, R, Rr,nl=2,nu=2);
#
# PDEs
   St=rep(0,nr);Iut=rep(0,nr);
  Iht=rep(0,nr); Rt=rep(0,nr);
#
# 0 le r le r0
  for(i in 1:nr){
#
# r=0
    if(i==1){
       St[i]=ds*2* Srr[i]-rate[i];
      Iut[i]=du*2*Iurr[i]+(1-mu)*rate[i]-ku*Iu[i];
      Iht[i]=dh*2*Ihrr[i]+mu*rate[i]-kh*Ih[i];
       Rt[i]=dr*2* Rrr[i]+ku*Iu[i]+kh*Ih[i];
    }
#
# r > 0
    if(i>1){
      ri=1/r[i];
       St[i]=ds*( Srr[i]+ri* Sr[i])-rate[i];
      Iut[i]=du*(Iurr[i]+ri*Iur[i])+(1-mu)*rate[i]-ku*Iu[i];
```

```
        Iht[i]=dh*(Ihrr[i]+ri*Ihr[i])+mu*rate[i]-kh*Ih[i];
         Rt[i]=dr*( Rrr[i]+ri* Rr[i])+ku*Iu[i]+kh*Ih[i];
     }
#
# Next i
   }
#
# Four vectors to one vector
   ut=rep(0,4*nr);
   for(i in 1:nr){
     ut[i]        =St[i];
     ut[i+nr]   =Iut[i];
     ut[i+2*nr]=Iht[i];
     ut[i+3*nr]= Rt[i];
   }
#
# Increment calls to pde1a
   ncall <<- ncall+1;
#
# Return derivative vector
   return(list(c(ut)));
   }
```

We can note the following details about Listing 2.2.

- The function is defined.

```
    pde1a=function(t,u,parms){
#
# Function pde1a computes the t derivative
# vectors of S(r,t),Iu(r,t),Ih(r,t),R(r,t)
```

t is the current value of t in Eqs. (1.1-1)–(1.1-4). u the 204-vector of ODE/MOL dependent variables. parm is an argument to pass parameters to pde1a (unused, but required in the argument list). The arguments must be listed in the order stated to properly interface with lsodes called in the main program of Listing 2.1. The derivative vector of the LHS of Eqs. (1.1-1)–(1.1-4) is calculated next and returned to lsodes.

- u is placed in four vectors, S,Iu,Ih,R, to facilitate the programming of Eqs. (1.2-1)–(1.2-4).

```
#
# One vector to four vectors
   S=rep(0,nr);Iu=rep(0,nr);
   Ih=rep(0,nr); R=rep(0,nr);
   for(i in 1:nr){
      S[i]=u[i];
      Iu[i]=u[i+nr];
      Ih[i]=u[i+2*nr];
      R[i]=u[i+3*nr];
   }
```

- The nonlinear algebra is computed for the RHSs of Eqs. (1.1-1)–(1.1-4), $\beta(I_u + \delta I_h)S$.

```
#
# Algebra
   rate=rep(0,nr);
   for(i in 1:nr){
      rate[i]=beta*(Iu[i]+delta*Ih[i])*S[i];
   }
```

- The derivatives $\dfrac{\partial S}{\partial r}, \dfrac{\partial I_u}{\partial r}, \dfrac{\partial I_h}{\partial r}, \dfrac{\partial R}{\partial r}$ in Eqs. (1.1-1)–(1.1-4) are computed by dss004.

```
#
# Sr,Iur,Ihr,Rr
   Sr=dss004(rl,ru,nr, S);
   Iur=dss004(rl,ru,nr,Iu);
   Ihr=dss004(rl,ru,nr,Ih);
   Rr=dss004(rl,ru,nr, R);
```

Sr,Iur,Ihr,Rr do not have to be allocated (with rep) since this is done by dss004. Additional details about dss004 are available in Appendix A.1.

- BCs (1.3) are implemented (the subscripts 1,nr correspond to $r = r_l, r_u$).

```
#
# BCs
   Sr[1]=0; Sr[nr]=0;
   Iur[1]=0;Iur[nr]=0;
   Ihr[1]=0;Ihr[nr]=0;
   Rr[1]=0; Rr[nr]=0;
```

- The derivatives $\dfrac{\partial^2 S}{\partial r^2}, \dfrac{\partial^2 I_u}{\partial r^2}, \dfrac{\partial^2 I_h}{\partial r^2}, \dfrac{\partial^2 R}{\partial r^2}$ in Eqs. (1.1-1)–(1.1-4) are computed by dss044. nl=2, nu=2 specify Neumann BCs and the BC values of the first derivatives are used in dss044.

```
#
# Srr,Iurr,Ihrr,Rrr
  nl=2;nu=2;
    Srr=dss044(rl,ru,nr, S, Sr,nl=2,nu=2);
    Iurr=dss044(rl,ru,nr,Iu,Iur,nl=2,nu=2);
    Ihrr=dss044(rl,ru,nr,Ih,Ihr,nl=2,nu=2);
    Rrr=dss044(rl,ru,nr, R, Rr,nl=2,nu=2);
```

Srr,Iurr,Ihrr,Rrr do not have to be allocated (with rep) since this is done by dss044. Additional details about dss044 are available in Appendix A.2.

- The LHS t derivatives of Eqs. (1.1-1)–(1.1-4), $\dfrac{\partial S}{\partial t}, \dfrac{\partial I_u}{\partial t}, \dfrac{\partial I_h}{\partial t}, \dfrac{\partial R}{\partial t}$, are allocated.

```
#
# PDEs
    St=rep(0,nr);Iut=rep(0,nr);
  Iht=rep(0,nr); Rt=rep(0,nr);
```

- The t derivatives in Eqs. (1.1-1)–(1.1-4) for $r_l \leq r \leq r_u$ are computed within a for.

```
#
# 0 le r le r0
  for(i in 1:nr){
```

- The four vectors St,Iut,Iht,Rt are computed according to Eqs. (1.1-1)–(1.1-4). This is done in two parts, for $r = 0, r > 0$ (because of the $\dfrac{1}{r}$ coefficient in Eqs. (1.1-1)–(1.1-4)).

```
#
# r=0
    if(i==1){
      St[i]=ds*2* Srr[i]-rate[i];
      Iut[i]=du*2*Iurr[i]+(1-mu)*rate[i]-ku*Iu[i];
      Iht[i]=dh*2*Ihrr[i]+mu*rate[i]-kh*Ih[i];
      Rt[i]=dr*2* Rrr[i]+ku*Iu[i]+kh*Ih[i];
    }
```

At $r = 0$ the radial groups in Eqs. (1.1-1)–(1.1-4) are indeterminant, and are resolved with l'Hospital's rule, e.g.,

$$\frac{1}{r}\frac{\partial S}{\partial r} = \frac{\partial^2 S}{\partial r^2}$$

and

$$\frac{\partial^2 S}{\partial r^2} + \frac{1}{r}\frac{\partial S}{\partial r} = 2\frac{\partial^2 S}{\partial r^2}.$$

BC (1.3-1) was used to evaluate the indeterminant term at $r = 0$ (with $r_l = 0$).

- For $r > 0$, Eqs. (1.1-1)–(1.1-4) are programmed as

```
#
# r > 0
    if(i>1){
      ri=1/r[i];
       St[i]=ds*( Srr[i]+ri* Sr[i])-rate[i];
       Iut[i]=du*(Iurr[i]+ri*Iur[i])+(1-mu)*rate[i]-ku*Iu[i];
       Iht[i]=dh*(Ihrr[i]+ri*Ihr[i])+mu*rate[i]-kh*Ih[i];
       Rt[i]=dr*( Rrr[i]+ri* Rr[i])+ku*Iu[i]+kh*Ih[i];
    }
```

The correspondence of Eqs. (1.1-1)–(1.1-4) and this coding demonstrates an important feature of the MOL. Specifically, the individual RHS functions and terms can be computed from the numerical solutions, for example, `beta*(Iu[i]+delta*Ih[i])*S[i]` from `S[i]`, `Iu[i]`, `Ih[i]`, then displayed numerically and graphically to elucidate the contributions to the PDE LHS t derivatives. This procedure is left as an exercise.

- The MOL over the grid in r is completed (with the `for`).

```
#
# Next i
    }
```

- The four vectors `St`,`Iut`,`Iht`,`Rt` are placed in a single derivative vector `ut` to return to `lsodes`.

```
#
# Four vectors to one vector
    ut=rep(0,4*nr);
    for(i in 1:nr){
       ut[i]       =St[i];
```

```
   ut[i+nr]  =Iut[i];
   ut[i+2*nr]=Iht[i];
   ut[i+3*nr]= Rt[i];
 }
```

- The counter for the calls to pde1a is incremented and returned to the main program of Listing 2.1 with <<-.

```
#
# Increment calls to pde1a
  ncall <<- ncall+1;
```

- ut is returned to lsodes as a list (required by lsodes). c is the R vector utility.

```
#
# Return derivative vector
  return(list(c(ut)));
 }
```

The final } concludes pde1a.

The output from the main program and subordinate routine of Listings 2.1, 2.2 is considered next.

2.3   MODEL OUTPUT

Abbreviated numerical output is in Table 2.1 for ncase=1.
    We can note the following details about this output.

- The dimensions of the solution matrix out are $nout \times 4nr + 1 = 16 \times 4(51) + 1 = 205$. The offset $+1$ results from the value of t as the first element in each of the $nout = 16$ solution vectors. These same values of t are in tout.

- ICs (1.2) are verified as programmed in the main program of Listing 2.1 for ncase=1, $S(r, t = 0) = 50$, $I_u(r, t = 0) = 0$, $I_h(r, t = 0) = 0$, $R(r, t = 0) = 0$.

- The output is for $r = 0, 50, 100$. In partcular, it is invariant in r

- The output is for $t = 0, 300$ as programmed in Listing 2.1. In particular, it is invariant in t.

Table 2.1: Abbreviated output for Eqs. (1.1-1)–(1.1-4), (1.2-1)–(1.2-4), and (1.3-1)–(1.3-8)
ncase=1

[1] 16

[1] 205

```
      t          r
               S(r,t)          Iu(r,t)
               Ih(r,t)          R(r,t)
      0      0.0
                  50.00              0.00
                   0.00              0.00
      0     50.0
                  50.00              0.00
                   0.00              0.00
      0    100.0
                  50.00              0.00
                   0.00              0.00


      t          r
               S(r,t)          Iu(r,t)
               Ih(r,t)          R(r,t)
    300      0.0
                  50.00              0.00
                   0.00              0.00
    300     50.0
                  50.00              0.00
                   0.00              0.00
    300    100.0
                  50.00              0.00
                   0.00              0.00
  ncall =      240
```

- $S(r, t)$, $I_u(r, t)$, $I_h(r, t)$, $R(x, t)$ remain at their initial values throughout the solution. This invariance with t follows from the zero t derivatives of Eqs. (1.1-1)–(1.1-4). For example, for Eq. (1.1-1), with constant derivatives in r, and BCs (1.3),

$$\frac{\partial S}{\partial t} = d_s \left(0 + \frac{1}{r}(0) \right) - \beta(0 + \delta(0))(50) = 0.$$

That is, all of the RHS terms of Eq. (1.1-1) are zero, and therefore the LHS t derivative is also zero.

The same conclusion follows for the t derivatives of Eqs. (1.1-2)–(1.1-4).

- The computational effort is modest, ncall = 240, as might be expected since the solutions do not change with t.

The graphical output confirms the invariance of the solutions in r and t. For example, Fig. 2.1-1 indicates that $S(r, t) = 50$.

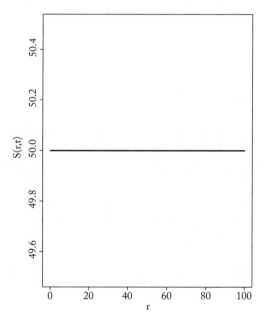

Figure 2.1-1: Numerical solution $S(r, t)$ from Eq. (1.1-1), ncase=1.

Similarly, Fig. 2.1-4 indicates that $R(r, t) = 0$.

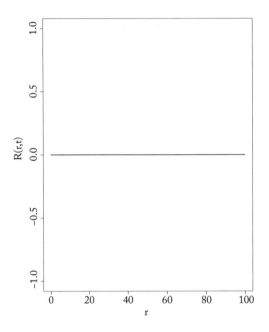

Figure 2.1-4: Numerical solution $R(r, t)$ from Eq. (1.1-4), ncase=1.

ncase=1 may seem trivial, but it is worth executing since if any of the PDE solutions depart from their initial values, a programming error would be indicated.

ncase=2 introduces a Gaussian IC for $I_u(r, t)$ which could, for example, represent the start of influenza near $r = 0$. In other words, the RHSs of Eqs. (1.1-1)–(1.1-4) are no longer zero, and the PDE solutions are a response to the Gaussian IC.

The numerical solution for ncase=2 appears in Table 2.2.

We can note the following details about this output.

- The dimensions of the solution matrix out are again $nout \times 4nr + 1 = 16 \times 4(51) + 1 = 205$ as expected, since the only changes from Table 2.1 to 2.2 is ncase=1 changed to ncase=2 in Listing 2.1.

- ICs (1.2) are verified as programmed in the main program of Listing 2.1 for ncase=2, $S(r, t = 0) = 50$, $I_u(r, t = 0) = 5e^{-0.01r^2}$, $I_h(r, t = 0) = 0$, $R(r, t = 0) = 0$.

- The output is for $r = 0, 50, 100$. In particular, it now varies with r.

- The output is for $t = 0, 300$ as programmed in Listing 2.1. In particular, it now varies with t.

Table 2.2: Abbreviated output for Eqs. (1.1-1)–(1.1-4), (1.2-1)–(1.2-4), and (1.3-1)–(1.3-8), ncase=2

[1] 16

[1] 205

t	r	S(r,t)	Iu(r,t)
		Ih(r,t)	R(r,t)
0	0.0		
		50.00	5.00
		0.00	0.00
0	50.0		
		50.00	0.00
		0.00	0.00
0	100.0		
		50.00	0.00
		0.00	0.00

t	r	S(r,t)	Iu(r,t)
		Ih(r,t)	R(r,t)
300	0.0		
		5.41	0.00
		0.00	44.97
300	50.0		
		5.58	0.00
		0.00	44.47
300	100.0		
		5.36	0.00
		0.00	44.64

ncall =   571

- $S(r,t)$, $I_u(r,t)$, $I_h(r,t)$, $R(x,t)$ depart from their initial values throughout the solution. That is, the RHS terms of Eqs. (1.1-1)–(1.1-4) are nonzero, and therefore the LHS t derivatives are also nonzero.

- An important property of the solution is that for large t (near 300) $S(r,t) + R(r,t) = 50$ with $S(r,t) \approx 5.5$ and $R(r,t) \approx 44.5$. Also, $I_u(r,t)$, $I_h(r,t)$ approach zero so that further infection ceases. These features can be observed in the graphical output, Figs. 2.2-1–2.2-4, that follow.

- The computational effort remains modest, `ncall` = 571, so that `lsodes` computes a solution efficiently for the 204 MOL ODEs.

The graphical output confirms the variation of the solutions in r and t.

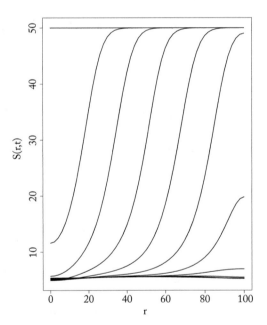

Figure 2.2-1: Numerical solution $S(r,t)$ from Eq. (1.1-1), `ncase=2`.

Figure 2.2-1 indicates that $S(r,t)$ starts from the IC $S(r, t = 0) = 50$, then evolves into a traveling wave moving left to right, until it reaches the right boundary. $S(r,t)$ approaches a steady state, uniform concentration of approximately 5.5 (Table 2.2).

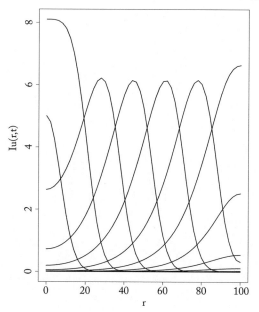

Figure 2.2-2: Numerical solution $I_u(r, t)$ from Eq. (1.1-2), `ncase=2`.

Figure 2.2-2 indicates that $I_u(r, t)$ starts from the IC $I_u(r, t = 0) = 5e^{-0.01r^2}$ (`ncase=2`, Listing 2.1), then evolves into a traveling wave (pulse) moving left to right, until it reaches the right boundary. $I_u(r, t)$ approaches a steady state, uniform concentration of approximately zero (Table 2.2).

Figure 2.2-3 indicates that $I_h(r, t)$ quickly responds to $I_u(r, t)$, then evolves into a traveling wave (pulse) moving left to right, until it reaches the right boundary. $I_h(r, t)$ approaches a steady state, uniform concentration of approximately zero (Table 2.2).

Figure 2.2-4 indicates that $R(r, t)$ starts from the IC $R(r, t = 0) = 0$, then evolves into a traveling wave moving left to right, until it reaches the right boundary. $R(r, t)$ approaches a steady state, uniform concentration of approximately 44.5 (Table 2.2).

In summary, most of the initial susceptibles eventually recover with no further influenza infection.

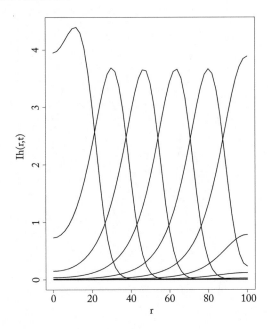

Figure 2.2-3: Numerical solution $I_h(r, t)$ from Eq. (1.1-3), ncase=2.

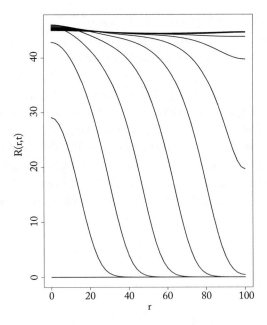

Figure 2.2-4: Numerical solution $R(r, t)$ from Eq. (1.1-4), ncase=2.

2.4 SUMMARY AND CONCLUSIONS

The influenza model of Eqs. (1.1-1)–(1.1-4), (1.2-1)–(1.2-4), and (1.3-1)–(1.3-8) is implemented within the MOL framework. For an intermediate level of treatment, $\mu = 0.5$, the susceptibles $S(r, t)$ and recovereds, $R(r, t)$, exhibit traveling front solutions, while the untreated and treated infecteds, $I_u(r, t)$, $Ih(r, t)$, exhibit traveling pulses with an eventual return to zero cases. The origin of these solution properties is studied further in the next two chapters.

REFERENCES

[1] Soetaert, K., Cash, J., and Mazzia, F. (2012). *Solving Differential Equations in R*, Springer-Verlag, Heidelberg, Germany. DOI: 10.1007/978-3-642-28070-2. 5

[2] Zhang, T. and Wang, W. (2014). Existence of traveling wave solutions for influenza model with treatment, *Journal of Mathematical Analysis and Applications*, 419, pp. 469–495. DOI: 10.1016/j.jmaa.2014.04.068. 9

CHAPTER 3

Model Analysis

INTRODUCTION

The PDE model formulation of Chapter 1 and the model computer implementation in Chapter 2 are now extended to a study of the model parameters and special case solutions. Also, 3D graphical (plotted) output is added to elucidate the solutions that are relatively complicated.

3.1   MAIN PROGRAM

The following Listing 3.1 contains a main program that is a minor variant of the main program of Listing 2.1.

Listing 3.1: Main program for Eqs. (1.1-1)–(1.1-4), (1.2-1)–(1.2-4), and (1.3-1)–(1.3-8)

```
#
# Influenza model
#
# Delete previous workspaces
  rm(list=ls(all=TRUE))
#
# Access ODE integrator
  library("deSolve");
#
# Access functions for numerical solution
  setwd("f:/flu/chap3");
  source("pde1a.R");
  source("dss004.R");
  source("dss044.R");
#
# Parameters for ODEs
  ncase=1;
  ds=1;
  du=1;
  dh=1;
  dr=1;
```

```
  beta=0.008;
  delta=0.6
  ku=0.1;
  kh=0.2;
  if(ncase==1){mu=0;}
  if(ncase==2){mu=1;}
  if(ncase==3){mu=0.5;}
#
# Spatial grid (in r)
  nr=51;
  rl=0;ru=100;
  r=seq(from=rl,to=ru,by=(ru-rl)/(nr-1));
#
# Independent variable for ODE integration
  t0=0;tf=300;nout=16;
  tout=seq(from=t0,to=tf,by=(tf-t0)/(nout-1));
#
# Initial condition (t=0)
  for(i in 1:nr){
    u0[i]      =50;
    u0[i+nr]   =5*exp(-0.01*r[i]^2);
    u0[i+2*nr]=0;
    u0[i+3*nr]=0;
  }
  ncall=0;
#
# ODE integration
  out=lsodes(y=u0,times=tout,func=pde1a,
      sparsetype="sparseint",rtol=1e-6,
      atol=1e-6,maxord=5);
  nrow(out)
  ncol(out)
#
# Arrays for plotting numerical solution
   S=matrix(0,nrow=nr,ncol=nout);
  Iu=matrix(0,nrow=nr,ncol=nout);
  Ih=matrix(0,nrow=nr,ncol=nout);
   R=matrix(0,nrow=nr,ncol=nout);
  for(it in 1:nout){
```

```
     for(i in 1:nr){
        S[i,it]=out[it,i+1];
       Iu[i,it]=out[it,i+1+nr];
       Ih[i,it]=out[it,i+1+2*nr];
        R[i,it]=out[it,i+1+3*nr];
     }
   }
#
# Display numerical solution
   cat(sprintf(" ncase = %2d, mu = %5.1f\n",ncase,mu));
   for(it in 1:nout){
     if((it-1)*(it-nout)==0){
       cat(sprintf("\n"));
       cat(sprintf("\n          t           r"));
       cat(sprintf("\n                        S(r,t)         Iu(r,t)"));
       cat(sprintf("\n                      Ih(r,t)          R(r,t)"));
     for(i in 1:nr){
       if((i-1)*(i-26)*(i-nr)==0){
       cat(sprintf("\n %6.0f %6.1f",tout[it],r[i]));
       cat(sprintf("\n          %12.2f %12.2f", S[i,it],Iu[i,it]));
       cat(sprintf("\n          %12.2f %12.2f",Ih[i,it], R[i,it]));
       }
     }
     }
   }
#
# Calls to ODE routine
   cat(sprintf("\n\n ncall = %5d\n\n",ncall));
#
# Plot PDE solutions
#
# S
   par(mfrow=c(1,1));
   matplot(r,S,type="l",xlab="r",ylab="S(r,t)",
     lty=1,main="",lwd=2,col="black");
#
# Iu
   par(mfrow=c(1,1));
   matplot(r,Iu,type="l",xlab="r",ylab="Iu(r,t)",
```

```
      lty=1,main="",lwd=2,col="black");
#
#  Ih
   par(mfrow=c(1,1));
   matplot(r,Ih,type="l",xlab="r",ylab="Ih(r,t)",
     lty=1,main="",lwd=2,col="black");
#
#  R
   par(mfrow=c(1,1));
   matplot(r,R,type="l",xlab="r",ylab="R(r,t)",
     lty=1,main="",lwd=2,col="black");
#
#  3D
#
#  S
   persp(r,tout,S,theta=120,phi=45,
         xlim=c(rl,ru),ylim=c(t0,tf),xlab="r",ylab="t",
         zlab="S(r,t)");
#
#  Iu
   persp(r,tout,Iu,theta=120,phi=45,
         xlim=c(rl,ru),ylim=c(t0,tf),xlab="r",ylab="t",
         zlab="Iu(r,t)");
#
#  Ih
   persp(r,tout,Ih,theta=120,phi=45,
         xlim=c(rl,ru),ylim=c(t0,tf),xlab="r",ylab="t",
         zlab="Ih(r,t)");
#
#  R
   persp(r,tout,R,theta=60,phi=45,
         xlim=c(rl,ru),ylim=c(t0,tf),xlab="r",ylab="t",
         zlab="R(r,t)");
```

We can note the following code added to Listing 2.1.

- Three cases are programmed corresponding to variations in μ in Eqs. (1.1-2) and (1.1-3).

```
   #
   # Parameters for ODEs
```

```
ncase=1;
ds=1;
du=1;
dh=1;
dr=1;
beta=0.008;
delta=0.6
ku=0.1;
kh=0.2;
if(ncase==1){mu=0;}
if(ncase==2){mu=1;}
if(ncase==3){mu=0.5;}
```

ncase=1 corresponds to no antiviral treatment of influenza.

- The same IC is used for the various values of ncase.

```
#
# Initial condition (t=0)
  for(i in 1:nr){
    u0[i]      =50;
    u0[i+nr]   =5*exp(-0.01*r[i]^2);
    u0[i+2*nr]=0;
    u0[i+3*nr]=0;
  }
  ncall=0;
```

In particular, as in Chapter 2, $S(r, t = 0) = 50$, $I_u(r, t = 0) = 5e^{-0.01r^2}$.

- 3D plotting with the R utility persp is added to the end of Listing 2.1 (after the 2D plotting with matplot).

```
#
# 3D
#
# S
  persp(r,tout,S,theta=120,phi=45,
        xlim=c(rl,ru),ylim=c(t0,tf),xlab="r",ylab="t",
        zlab="S(r,t)");
#
# Iu
```

```
     persp(r,tout,Iu,theta=120,phi=45,
          xlim=c(rl,ru),ylim=c(t0,tf),xlab="r",ylab="t",
          zlab="Iu(r,t)");
  #
  # Ih
     persp(r,tout,Ih,theta=120,phi=45,
          xlim=c(rl,ru),ylim=c(t0,tf),xlab="r",ylab="t",
          zlab="Ih(r,t)");
  #
  # R
     persp(r,tout,R,theta=60,phi=45,
          xlim=c(rl,ru),ylim=c(t0,tf),xlab="r",ylab="t",
          zlab="R(r,t)");
```

In general, the four PDE dependent variables are plotted against r and t in 3D as reflected in figures discussed subsequently.

The MOL/ODE routine pde1a called by lsodes is the same as in Listing 2.2 so it is not repeated here. The output for ncase=1,2,3 is considered next.

3.2   MODEL OUTPUT, NO TREATMENT

The output from the main program and ODE/MOL routine of Listings 3.1, 2.2 follows for ncase=1. See Table 3.1.

We can note the following details about this output.

- The dimensions of the solution matrix out are $nout \times 4nr + 1 = 16 \times 4(51) + 1 = 205$. The offset $+1$ results from the value of t as the first element in each of the $nout = 16$ solution vectors. These same values of t are in tout.

- ICs (1.2) are verified as programmed in the main program of Listing 3.1, $S(r, t = 0) = 50$, $I_u(r, t = 0) = 5e^{-0.01r^2}$, $I_h(r, t = 0) = R(r, t = 0) = 0$.

- The output is for $r = 0, 50, 100$ as programmed in Listing 3.1.

- The output is for $t = 0, 300$ as programmed in Listing 3.1.

- $S(r, t)$, $I_u(r, t)$, $R(x, t)$ exhibit moving front solutions that are elucidated by the 2D and 3D plots that follow. $I_h(r, t)$ remains at the zero IC for no treatment, $\mu = 0$ (all of the RHS terms of Eq. (1.1-3) remain at zero). Also, $S(r, t) + I_u(r, t) + I_h(r, t) + R(r, t) = 50$ for large t as explained subsequently.

- The computational effort is modest, ncall = 633 so that lsodes efficiently computes a solution to the 204 MOL/ODEs.

Table 3.1: Abbreviated output for Eqs. (1.1-1)–(1.1-4), (1.2-1)–(1.2-4), and (1.3-1)–(1.3-8), ncase=1

```
[1] 16

[1] 205

ncase =  1, mu =    0.0

     t        r
             S(r,t)       Iu(r,t)
             Ih(r,t)       R(r,t)
     0    0.0
                50.00         5.00
                 0.00         0.00
     0   50.0
                50.00         0.00
                 0.00         0.00
     0  100.0
                50.00         0.00
                 0.00         0.00


     t        r
             S(r,t)       Iu(r,t)
             Ih(r,t)       R(r,t)
   300    0.0
                 1.24         0.00
                 0.00        49.15
   300   50.0
                 1.31         0.00
                 0.00        48.75
   300  100.0
                 1.24         0.00
                 0.00        48.76

ncall =     633
```

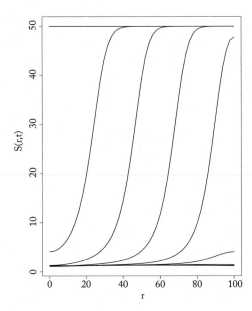

Figure 3.1-1: Numerical solution $S(r,t)$ from Eq. (1.1-1), `ncase=1`.

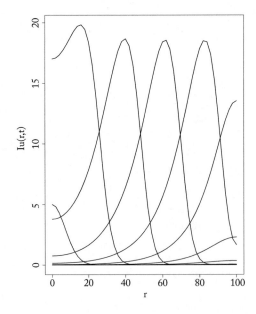

Figure 3.1-2: Numerical solution $I_u(r,t)$ from Eq. (1.1-2), `ncase=1`.

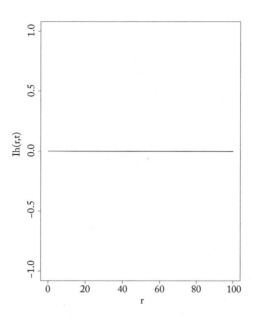

Figure 3.1-3: Numerical solution $I_h(r, t)$ from Eq. (1.1-3), ncase=1.

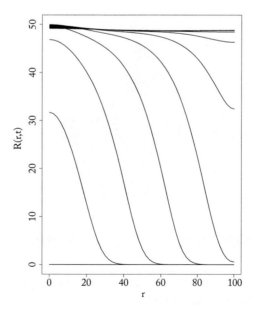

Figure 3.1-4: Numerical solution $R(r, t)$ from Eq. (1.1-4), ncase=1.

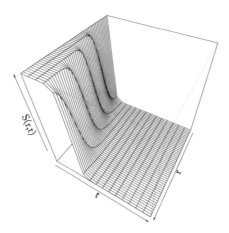

Figure 3.1-5: Numerical solution $S(r,t)$ from Eq. (1.1-1), `ncase=1`.

The graphical output displays the moving front solutions.

Figures 3.1-1 and 3.1-5 (2D,3D) indicate the reduction in susceptibles from $S(r, t = 0) = 50$ to a final level of approximately 1.24 (Table 3.1) as a moving front.

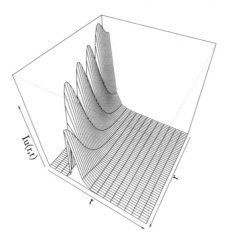

Figure 3.1-6: Numerical solution $I_u(r,t)$ from Eq. (1.1-2), `ncase=1`.

Figures 3.1-2 and 3.1-6 (2D,3D) indicate the variation in untreated cases, $I_u(r,t)$, as a moving pulse with a final value of zero (Table 3.1).

Figure 3.1-3 (2D) indicates that the treated cases, $I_h(r,t)$, remain at the zero initial value for $\mu = 0$, as expected with no treatment. `persp` did not produce a 3D plot because the z axis has no variation in r or t.

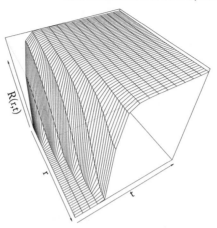

Figure 3.1-8: Numerical solution $R(r, t)$ from Eq. (1.1-4), `ncase=1`.

Figures 3.1-4 and 3.1-8 (2D,3D) indicate an increase in recoverds, $R(r, t)$, from the zero initial condition to a final level of approximately 48.7 (Table 3.1) as a moving front.

In summary, most of the initial susceptibles eventually become recovereds with no treatment ($\mu = 0$) so that the influenza outbreak is stable. This result suggests some revision of the model to reflect deaths of susceptibles from influenza and a reduction in recovereds with no treatment.

3.3   MODEL OUTPUT, COMPLETE TREATMENT

The output from the main program and ODE/MOL routine of Listings 3.1, 2.2 follows for `ncase=2`. See Table 3.2.

We can note the following details about this output.

- The dimensions of the solution matrix `out` are again $nout \times 4nr + 1 = 16 \times 4(51) + 1 = 205$.

- ICs (1.2) are verified as programmed in the main program of Listing 3.1, $S(r, t = 0) = 50$, $I_u(r, t = 0) = 5e^{-0.01r^2}$, $I_h(r, t = 0) = R(r, t = 0) = 0$.

- The output is for $r = 0, 50, 100$ as programmed in Listing 3.1.

- The output is for $t = 0, 300$ as programmed in Listing 3.1.

- $S(r, t)$, $I_u(r, t)$, $I_h(r, t)$, $R(x, t)$ exhibit moving front solutions that are elucidated by the 2D and 3D plots that follow. $I_h(r, t)$ does not remain at the zero IC with complete treatment ($\mu = 1$). Also, $S(r, t) + I_u(r, t) + I_h(r, t) + R(r, t) = 50$ as explained subsequently.

Table 3.2: Abbreviated output for Eqs. (1.1-1)–(1.1-4), (1.2-1)–(1.2-4), and (1.3-1)–(1.3-8), ncase=2

```
[1] 16

[1] 205

ncase =  2, mu =   1.0


    t        r
              S(r,t)        Iu(r,t)
              Ih(r,t)        R(r,t)
    0     0.0
              50.00          5.00
               0.00          0.00
    0    50.0
              50.00          0.00
               0.00          0.00
    0   100.0
              50.00          0.00
               0.00          0.00


    t        r
              S(r,t)        Iu(r,t)
              Ih(r,t)        R(r,t)
  300     0.0
              34.15         -0.00
               0.00         16.24
  300    50.0
              35.93          0.00
               0.06         14.06
  300   100.0
              38.75          0.00
               0.56         10.69


  ncall =    433
```

• The computational effort is modest, `ncall` = 433 so that `lsodes` efficiently computes a solution to the 204 MOL/ODEs.

The graphical output displays the moving front solutions.

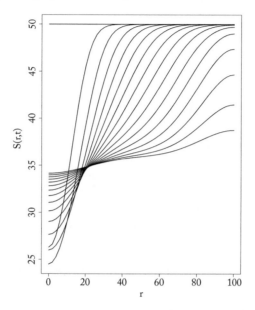

Figure 3.2-1: Numerical solution $S(r,t)$ from Eq. (1.1-1), `ncase=2`.

Figures 3.2-1 and 3.2-5 (2D,3D) indicate the reduction in susceptibles from $S(r,t=0) = 50$ to a final level that varies from 34.15 to 38.75 (Table 3.2) as a moving front.

Figures 3.2-2 and 3.2-6 (2D,3D) indicate the variation in untreated cases, $I_u(r,t)$, as a moving pulse with a final level of zero (Table 3.2).

Figures 3.2-3 and 3.2-7 (2D,3D) indicate the variation in treated cases, $I_h(r,t)$, as a moving pulse with a final value that varies from 0 to 0.56 (Table 3.2).

Figures 3.2-4 and 3.2-8 (2D,3D) indicate an increase in recoverds, $R(r,t)$, from the zero initial condition to a final level that varies from 16.24 to 10.69 (Table 3.2) as a moving front.

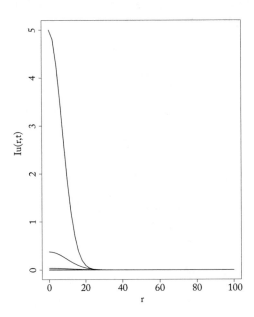

Figure 3.2-2: Numerical solution $I_u(r, t)$ from Eq. (1.1-2), ncase=2.

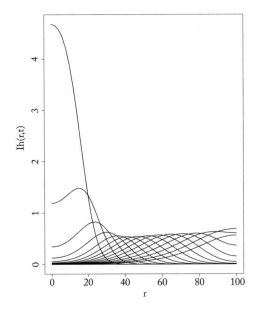

Figure 3.2-3: Numerical solution $I_h(r, t)$ from Eq. (1.1-3), ncase=2.

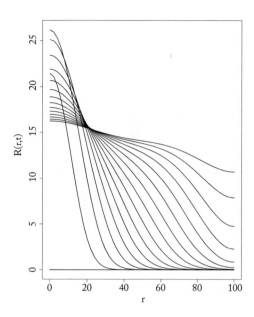

Figure 3.2-4: Numerical solution $R(r, t)$ from Eq. (1.1-4), ncase=2.

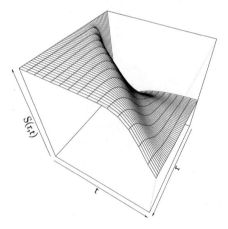

Figure 3.2-5: Numerical solution $S(r, t)$ from Eq. (1.1-1), ncase=2.

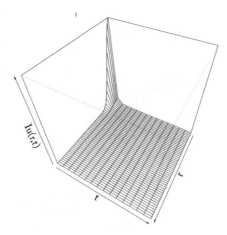

Figure 3.2-6: Numerical solution $I_u(r, t)$ from Eq. (1.1-2), ncase=2.

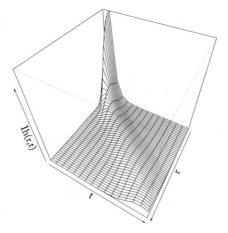

Figure 3.2-7: Numerical solution $I_h(r, t)$ from Eq. (1.1-2), ncase=2.

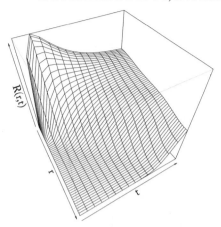

Figure 3.2-8: Numerical solution $R(r, t)$ from Eq. (1.1-4), `ncase=2`.

In summary, for `ncase=2`, the reduction in the susceptibles is mitigated by the antiviral treatment (comparing $S(r, t)$ values in Tables 3.1 and 3.2). Also, the increase in recovereds for `ncase=1` (Table 3.1) is reduced for `ncase=2` (Table 3.2) since the treatment reduces the cases of influenza infection that eventually lead to recovereds.

To conclude the study of treatment levels, `ncase=3` is for an intermediate level of treatment, $\mu = 0.5$ (Listing 3.1). This is also the `ncase=2` of Listing 2.1, but the numerical output is repeated here (Tables 2.2 and 3.3 are the same), and 3D graphical output is presented here for `ncase=3` (the 2D graphical output is in Figures 2.2-1–2.2-4).

3.4 MODEL OUTPUT, INTERMEDIATE TREATMENT

The output from the main program and ODE/MOL routine of Listings 3.1, 2.2 follows for `ncase=3`. See Table 3.3.

We can note the following details about this output.

- The dimensions of the solution matrix out are again $nout \times 4nr + 1 = 16 \times 4(51) + 1 = 205$.

- ICs (1.2) are verified as programmed in the main program of Listing 3.1, $S(r, t = 0) = 50$, $I_u(r, t = 0) = 5e^{-0.01r^2}$, $I_h(r, t = 0) = R(r, t = 0) = 0$.

- The output is for $r = 0, 50, 100$ as programmed in Listing 3.1.

- The output is for $t = 0, 300$ as programmed in Listing 3.1.

- $S(r, t)$, $I_u(r, t)$, $I_h(r, t)$, $R(x, t)$ exhibit moving front solutions that are elucidated by the 3D plots which follow (the 2D plots are in Figs. 2.2-1–2.2-4). (2.2)). Both $I_u(r, t)$ and

Table 3.3: Abbreviated output for Eqs. (1.1-1)–(1.1-4), (1.2-1)–(1.2-4), and (1.3-1)–(1.3-8), ncase=3

[1] 16

[1] 205

ncase = 3, mu =   0.5

t	r	S(r,t)	Iu(r,t)
		Ih(r,t)	R(r,t)
0	0.0		
		50.00	5.00
		0.00	0.00
0	50.0		
		50.00	0.00
		0.00	0.00
0	100.0		
		50.00	0.00
		0.00	0.00

t	r	S(r,t)	Iu(r,t)
		Ih(r,t)	R(r,t)
300	0.0		
		5.41	0.00
		0.00	44.97
300	50.0		
		5.58	0.00
		0.00	44.47
300	100.0		
		5.36	0.00
		0.00	44.64

ncall =   571

$I_h(r, t)$ depart from the zero ICs. Also, $S(r, t) + I_u(r, t) + I_h(r, t) + R(r, t) = 50$ as explained subsequently.

- The computational effort is modest, `ncall = 571` so that `lsodes` efficiently computes a solution to the 204 MOL/ODEs.

The following graphical output (3D only) displays the moving front solutions.

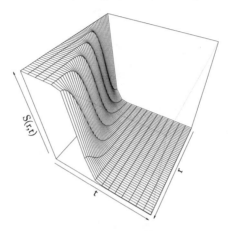

Figure 3.3-5: Numerical solution $S(r, t)$ from Eq. (1.1-1), `ncase=3`.

Figures 3.2-1 and 3.3-5 (2D,3D) indicate the reduction in susceptibles from $S(r, t = 0) = 50$ to a final level that varies from 5.41 to 5.36 (Table 3.3) as a moving front).

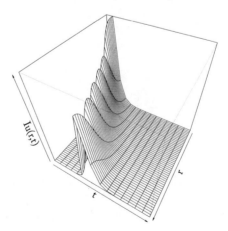

Figure 3.3-6: Numerical solution $I_u(r, t)$ from Eq. (1.1-2), `ncase=3`.

Figures 3.2-2 and 3.3-6 (2D,3D) indicate the variation in untreated cases, $I_u(r, t)$, as a moving pulse with a final level of zero (Table 3.3).

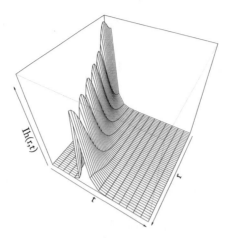

Figure 3.3-7: Numerical solution $I_h(r, t)$ from Eq. (1.1-2), ncase=3.

Figures 3.2-3 and 3.3-7 (2D,3D) indicate the variation in treated cases, $I_h(r, t)$, as a moving pulse with a final level of zero (Table 3.3).

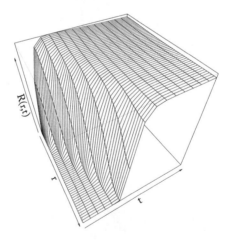

Figure 3.3-8: Numerical solution $R(r, t)$ from Eq. (1.1-4), ncase=3.

Figures 3.2-4 and 3.3-8 (2D,3D) indicate an increase in recoverds, $R(r, t)$, from the zero initial condition to a final level that varies from 44.97 to 44.64 (Table 3.3) as a moving front.

For ncase=3, the reduction in the suceptibles is intermediate between the values for ncase=1,2. For example, see Table 3.4 for $t = 300$, $r = 50$.

Table 3.4: Comparison of solutions for ncase=1,2,3

```
ncase=1, mu=0 (Table 3.1)

   300   100.0
               1.24              0.00
               0.00             48.76

ncase=2, mu=1 (Table 3.2)

   300   100.0
              38.75              0.00
               0.56             10.69

ncase=3, mu=0.5 (Table 3.3)

   300   100.0
               5.36              0.00
               0.00             44.64
```

In summary, the susceptibles who become infected drop, and the recovereds therefore drop, with increasing treatment. In each case, the sum of the four PDE dependent variables is a constant as explained next.

3.5 TOTAL POPULATION ANALYSIS

To analyze the constant total population, Eqs. (1.1-1)–(1.1-4) are summed

$$\frac{\partial S}{\partial t} + \frac{\partial I_u}{\partial t} + \frac{\partial I_h}{\partial t} + \frac{\partial R}{\partial t} =$$

$$d_s \left(\frac{\partial^2 S}{\partial r^2} + \frac{1}{r}\frac{\partial S}{\partial r} \right) +$$

$$d_u \left(\frac{\partial^2 I_u}{\partial r^2} + \frac{1}{r}\frac{\partial I_u}{\partial r} \right) +$$

$$d_h \left(\frac{\partial^2 I_h}{\partial r^2} + \frac{1}{r}\frac{\partial I_h}{\partial r} \right) +$$

$$d_r \left(\frac{\partial^2 R}{\partial r^2} + \frac{1}{r}\frac{\partial R}{\partial r} \right).$$

Note that the RHS algebraic terms cancel which is a basic requirement for the populations to sum to a constant. The preceding equation with the same value of the diffusivities, $d_s = d_u = d_h = d_r = d_{sr}$, is

$$\frac{\partial (S + I_u + I_h + R)}{\partial t} =$$

$$d_{sr} \left(\frac{\partial^2 (S + I_u + I_h + R)}{\partial r^2} + \frac{1}{r} \frac{\partial (S + I_u + I_h + R)}{\partial r} \right)$$

or, if $S + I_u + I_h + R = N$ where N is the total constant population (no variation with r or t)

$$\frac{\partial N}{\partial t} =$$

$$d_{sr} \left(\frac{\partial^2 N}{\partial r^2} + \frac{1}{r} \frac{\partial N}{\partial r} \right)$$

or

$$0 = 0$$

as hypothesized.

In summary, if the PDE model is altered (for example, during model formulation) so that the RHS algebraic variables do not sum to zero, this constant constraint on the total population will not apply. However, this case (RHS algebraic variables sum to zero) is still worth considering as a check on the programming, particularly in pde1a of Listing 2.2.

3.6   SUMMARY AND CONCLUSIONS

The preceding numerical analysis of the solution of the PDE model of Eqs. (1.1-1)–(1.1-4), (1.2-1)–(1.2-4), and (1.3-1)–(1.3-8) focuses on the variation of the basic parameter μ that defines the level of treatment. This analysis demonstrates how a model can be studied experimentally on a computer, including possibly a special case check such as the constant total population constraint. This type of checking (error analysis) is important, especially when an analytical check is not available, which is usually the case in realistic applications.

In the next chapter, alternative programming of the PDE model, particularly for pde1a, is considered.

CHAPTER 4

Moving Boundary Model

INTRODUCTION

To conclude the discussion of the influenza model of Eqs. (1.1-1)–(1.1-4), (1.2-1)–(1.2-4), and (1.3-1)–(1.3-8), the movement of the outer boundary at $r = r_u$ is considered. That is, the region of influenza infection can change with evolving populations.

The movement of the outer boundary is defined according to a velocity equation. The algorithm for the moving boundary partial differential equation (MBPDE) model is implemented as an extension of the main program of Listing 3.1.

4.1 MAIN PROGRAM

The MBPDE main program appears in Listing 4.1.

Listing 4.1: Main program for MBPDE form of Eqs. (1.1-1)–(1.1-4), (1.2-1)–(1.2-4), and (1.3-1)–(1.3-8)

```
#
#  Four MBPDE model
#
# Delete previous workspaces
  rm(list=ls(all=TRUE))
#
# Access ODE integrator
  library("deSolve");
#
# Access functions for numerical solution
  setwd("f:/flu/chap4");
  source("pde1a.R");
  source("dss004.R");
  source("dss044.R");
#
# Select case
  ncase=1;
#
```

```
# Parameters for ODEs
  ds=1;
  du=1;
  dh=1;
  dr=1;
  beta=0.008;
  mu=0.5;
  delta=0.6
  ku=0.1;
  kh=0.2;
#
# Spatial grid (in r)
  nr=51;rl=0;ru=100;dr=(ru-rl)/(nr-1);
  r=seq(from=rl,to=ru,by=dr);
#
# Independent variable for ODE integration
  t0=0;tf=300;
  tout=rep(0,2);
  nout=16;dt=(tf-t0)/(nout-1);it=1;
  tp=rep(0,nout);rup=rep(0,nout);
  drdtp=rep(0,(nout-1));
#
# Initial condition (t=0)
  u0=rep(0,4*nr);
  for(i in 1:nr){
        u0[i]=50;                    #  S(r,t=0)
      u0[i+nr]=5*exp(-0.01*r[i]^2); # Iu(r,t=0)
    u0[i+2*nr]=0;                    # Ih(r,t=0)
    u0[i+3*nr]=0;                    #  R(r,t=0)
  }
#
# Arrays for MBPDE algorithm
    S=rep(0,nr);
   Iu=rep(0,nr);
   Ih=rep(0,nr);
    R=rep(0,nr);
#
# First point (IC) for MBPDE algorithm
  for(i in 1:nr){
```

```
       S[i]=      u0[i];
      Iu[i]=   u0[i+nr];
      Ih[i]=u0[i+2*nr];
       R[i]=u0[i+3*nr];
  }
#
# Arrays for plotting
    Sp=matrix(0,nrow=nr,ncol=nout);
   Iup=matrix(0,nrow=nr,ncol=nout);
   Ihp=matrix(0,nrow=nr,ncol=nout);
    Rp=matrix(0,nrow=nr,ncol=nout);
#
# First plotted point
  for(i in 1:nr){
      Sp[i,1]= S[i];
     Iup[i,1]=Iu[i];
     Ihp[i,1]=Ih[i];
      Rp[i,1]= R[i];
  }
  tp[1]=t0;rup[1]=ru;
  ncall=0;
#
# Next step along solution
  while(it<nout){
  for(i in 1:nr){
         u0[i]= S[i];
      u0[i+nr]=Iu[i];
    u0[i+2*nr]=Ih[i];
    u0[i+3*nr]= R[i];
  }
  t0=tout[2];
  tout[1]=t0;
  tout[2]=tout[1]+dt;
#
# ODE integration
  out=lsodes(y=u0,times=tout,func=pde1a,
      sparsetype="sparseint",rtol=1e-10,
      atol=1e-10,maxord=5);
#
```

```
# Arrays for solution
  for(i in 1:nr){
    S[i]=      out[2,i+1];
   Iu[i]=   out[2,i+1+nr];
   Ih[i]=out[2,i+1+2*nr];
    R[i]=out[2,i+1+3*nr];
  }
#
# Redefine spatial grid
    tableS  =splinefun(r,S);
   tableIu =splinefun(r,Iu);
   tableIh =splinefun(r,Ih);
    tableR  =splinefun(r,R);
  if(ncase==1){drdt=0;}
  if(ncase==2){drdt=0.2;}
  if(ncase==3){drdt=0.005*S[nr];}
  ru=ru+drdt*dt;
  r=seq(from=rl,to=ru,by=(ru-rl)/(nr-1));
#
# Solution on redefined grid
    S  =tableS(r,deriv=0);
   Iu =tableIu(r,deriv=0);
   Ih =tableIh(r,deriv=0);
    R  =tableR(r,deriv=0);
  it=it+1;
  rup[it]=ru;drdtp[it-1]=drdt;
#
# Solution for plotting
  for(i in 1:nr){
     Sp[i,it]= S[i];
    Iup[i,it]=Iu[i];
    Ihp[i,it]=Ih[i];
     Rp[i,it]= R[i];
  }
  tp[it]=tout[2];
#
# Next step (from while)
  }
#
```

```
# Display numerical solution
  cat(sprintf(" ncase = %2d, mu = %4.1f\n",ncase,mu));
  for(it in 1:nout){
    if((it-1)*(it-nout)==0){
      cat(sprintf("\n"));
      cat(sprintf("\n        t         r"));
      cat(sprintf("\n                  S(r,t)      Iu(r,t)"));
      cat(sprintf("\n                  Ih(r,t)       R(r,t)"));
    for(i in 1:nr){
      if((i-1)*(i-26)*(i-nr)==0){
      cat(sprintf("\n %6.0f %6.1f",tp[it],r[i]));
      cat(sprintf("\n            %12.2f %12.2f", Sp[i,it],Iup[i,it]))
          ;
      cat(sprintf("\n            %12.2f %12.2f",Ihp[i,it], Rp[i,it]))
          ;
      }
    }
    }
  }
#
# Plot PDE solutions
#
# 2D
#
# S
  par(mfrow=c(1,1));
  matplot(r,Sp,type="l",xlab="r",ylab="S(r,t)",
    lty=1,main="",lwd=2,col="black");
#
# Iu
  par(mfrow=c(1,1));
  matplot(r,Iup,type="l",xlab="r",ylab="Iu(r,t)",
    lty=1,main="",lwd=2,col="black");
#
# Ih
  par(mfrow=c(1,1));
  matplot(r,Ihp,type="l",xlab="r",ylab="Ih(r,t)",
    lty=1,main="",lwd=2,col="black");
#
```

```
# R
  par(mfrow=c(1,1));
  matplot(r,Rp,type="l",xlab="r",ylab="R(r,t)",
    lty=1,main="",lwd=2,col="black");
#
# 3D
#
# S
  persp(r,tp,Sp,theta=120,phi=45,xlab="r",ylab="t",
        zlab="S(r,t)");
#
# Iu
  persp(r,tp,Iup,theta=120,phi=45,xlab="r",ylab="t",
        zlab="Iu(r,t)");
#
# Ih
  persp(r,tp,Ihp,theta=120,phi=45,xlab="r",ylab="t",
        zlab="Ih(r,t)");
#
# R
  persp(r,tp,Rp,theta=60,phi=45,xlab="r",ylab="t",
        zlab="R(r,t)");
#
# Boundary position
  plot(tp,rup,xlab="t",ylab="ru");
    lines(tp,rup,type="l",lwd=2);
  points(tp,rup,pch="o",lwd=2);
  plot(tp[2:nout],drdtp,xlab="t",ylab="dru/dt");
    lines(tp[2:nout],drdtp,type="l",lwd=2);
  points(tp[2:nout],drdtp,pch="o",lwd=2);
#
# Calls to ODE routine
  cat(sprintf("\n\n  ncall = %3d\n",ncall));
```

This main program has substantial additions to the main program of Listing 3.1, so it is considered in detail with some repetition of the preceding discussion.

- Previous workspaces are deleted.

```
    #
```

```
#   Four MBPDE model
#
# Delete previous workspaces
  rm(list=ls(all=TRUE))
```

- The R ODE integrator library deSolve is accessed. Then the directory with the files for the solution of Eqs. (1.1-1)–(1.1-4), (1.2-1)–(1.2-4), and (1.3-1)–(1.3-8) is designated. Note that setwd (set working directory) uses / rather than the usual \.

```
#
# Access ODE integrator
  library("deSolve");
#
# Access functions for numerical solution
  setwd("f:/flu/chap4");
  source("pde1a.R");
  source("dss004.R");
  source("dss044.R");
```

The MOL/ODE routine, pde_1a.R, is the same as in Listing 2.2. In other words, the MF-PDE algorithm is confined to the main program and the discussion of Listing 2.2 serves as some of the explanation here. dss004, dss044 are library routines for the calculation of first and second order spatial derivatives in pde1a.

- ncase is specified with three possible values, ncase=1,2,3, corresponding to different velocities for the moving boundary at $r = r_u$.

```
#
# Select case
  ncase=1;
```

- The parameters are specified (as in Listings 2.1, 3.1). In particular, $\mu = 0.5$.

```
#
# Parameters for ODEs
  ds=1;
  du=1;
  dh=1;
  dr=1;
  beta=0.008;
  mu=0.5;
```

```
delta=0.6
ku=0.1;
kh=0.2;
```

- A spatial grid of 51 points is defined for $r_l = 0 \le r \le r_u = 100$ so that $r = 0, 2, \ldots, 100$.

```
#
# Spatial grid (in r)
  nr=51;
  rl=0;ru=100;
  r=seq(from=rl,to=ru,by=(ru-rl)/(nr-1));
```

- Parameters for the MOL solution are defined.

```
#
# Independent variable for ODE integration
  t0=0;tf=300;
  tout=rep(0,2);
  nout=16;dt=(tf-t0)/(nout-1);it=1;
  tp=rep(0,nout);rup=rep(0,nout);
  drdtp=rep(0,(nout-1));
```

These statements require some additional explanation.

- The initial and final values of t for the solution are defined.

    ```
    t0=0;tf=300;
    ```

- Rather than call lsodes once to compute a complete solution from t_0 to a final time t_f, lsodes is called for a series of output points in vector tout. In each of these intervals of two points, the grid in r is defined for an updated r_u (calculated by Euler's method, ru=ru+drdt*dt). In this way, r_u is refined as the solution proceeds to reflect the moving boundary.

    ```
    tout=rep(0,2);
    ```

- Each interval of two points has length dt=(300-0)/(16-1)=20 and 16 output points are defined, nout=16 (including $t = t_0$). The first point in each interval has index it=1 and the second point has index it=2.

    ```
    nout=16;dt=(tf-t0)/(nout-1);it=1;
    ```

- The value of t at the 16 output points is placed in vector tp and the corresponding values of r_u are placed in vector rup for plotting. In this way, the movement of the boundary at $r = r_u$ can be observed graphically.

```
tp=rep(0,nout);rup=rep(0,nout);
```

– Similarly, the varying values of $\dfrac{dr_u}{dt}$ (for ncase=1,2,3) are placed in drdtp for plotting. Since this derivative is not available initially at $t = t_0$ (but only after ru=ru+drdt*dt is used), there are $16 - 1 = 15$ values of the derivative.

```
drdtp=rep(0,(nout-1));
```

- The IC vectors for $S(r, t = 0)$, $I_u(r, t = 0)$, $I_h(r, t = 0)$, $R(r, t = 0)$ are defined numerically.

```
#
# Initial condition (t=0)
  u0=rep(0,4*nr);
  for(i in 1:nr){
        u0[i]=50;                      #  S(r,t=0)
      u0[i+nr]=5*exp(-0.01*r[i]^2);  # Iu(r,t=0)
    u0[i+2*nr]=0;                     # Ih(r,t=0)
    u0[i+3*nr]=0;                     #  R(r,t=0)
  }
```

The four IC vectors are placed in a single vector u0 for the start of the integration of the 4*nr = 4*51 = 204 MOL ODEs.

- The IC vectors are placed in S,Iu,Ih,R for subsequent use in the moving boundary algorithm and in Sp,Iup,Ihp,Rp for plotting. The initial values t = tp[1] and r_u = rup[1] are also defined.

```
#
# Arrays for MBPDE algorithm
    S=rep(0,nr);
   Iu=rep(0,nr);
   Ih=rep(0,nr);
    R=rep(0,nr);
#
# First point (IC) for MBPDE algorithm
  for(i in 1:nr){
       S[i]=      u0[i];
      Iu[i]=  u0[i+nr];
      Ih[i]=u0[i+2*nr];
       R[i]=u0[i+3*nr];
  }
```

```
#
# Arrays for plotting
    Sp=matrix(0,nrow=nr,ncol=nout);
   Iup=matrix(0,nrow=nr,ncol=nout);
   Ihp=matrix(0,nrow=nr,ncol=nout);
    Rp=matrix(0,nrow=nr,ncol=nout);
#
# First plotted point
  for(i in 1:nr){
      Sp[i,1]= S[i];
     Iup[i,1]=Iu[i];
     Ihp[i,1]=Ih[i];
      Rp[i,1]= R[i];
  }
  tp[1]=t0;rup[1]=ru;
  ncall=0;
```

The counter for the calls to the ODE/MOL routine pde_1a is initialized.

- The next interval of two points is initialized. For it=1, the first interval $0 \leq t \leq dt = (300 - 0)/(16 - 1) = 20$ is initialized (using the while).

```
#
# Next step along solution
  while(it<nout){
  for(i in 1:nr){
         u0[i]= S[i];
      u0[i+nr]=Iu[i];
    u0[i+2*nr]=Ih[i];
    u0[i+3*nr]= R[i];
  }
  t0=tout[2];
  tout[1]=t0;
  tout[2]=tout[1]+dt;
```

- The system of $4(51) = 204$ MOL/ODEs is integrated by the library integrator lsodes (available in deSolve [1]). As expected, the inputs to lsodes are the ODE function, pde_1a, the IC vector u0, and the vector of output values of t, tout. The length of u0 ($4(51) = 204$) informs lsodes how many ODEs are to be integrated. func,y,times are reserved names.

```
#
# ODE integration
  out=lsodes(y=u0,times=tout,func=pde1a,
      sparsetype="sparseint",rtol=1e-10,
      atol=1e-10,maxord=5);
```

The numerical solution to the ODEs is returned in matrix out. In this case, out has the dimensions $2 \times 4nr + 1 = 2 \times 4(51) + 1 = 205$.

The offset $+ 1$ is required since the first element of each column has the output t (also in tout), and the $2, \ldots, 4nr + 1 = 2, \ldots, 205$ column elements have the 204 ODE solutions.

- The solution of the 204 ODEs returned in out by lsodes is placed in vectors S,Iu,Ih,R of length 51 (defined previously).

```
#
# Arrays for solution
  for(i in 1:nr){
    S[i]=      out[2,i+1];
    Iu[i]=  out[2,i+1+nr];
    Ih[i]=out[2,i+1+2*nr];
    R[i]=out[2,i+1+3*nr];
  }
```

The solutions from out each have two points in t. The offset i+1 is required since the first element in each solution vector has the value of t and the 2 to 4nr+1 = 205 elements have the solutions of Eqs. (1.1-1)–(1.1-4).

- ru=ru+drdt*dt (Euler's method) is used to redefine r_u, where $drdt$ is the outer boundary velocity defined for three cases, ncase=1,2,3.

```
#
# Redefine spatial grid
    tableS  =splinefun(r,S);
   tableIu =splinefun(r,Iu);
   tableIh =splinefun(r,Ih);
    tableR  =splinefun(r,R);
  if(ncase==1){drdt=0;}
  if(ncase==2){drdt=0.2;}
  if(ncase==3){drdt=0.005*S[nr];}
  ru=ru+drdt*dt;
  r=seq(from=rl,to=ru,by=(ru-rl)/(nr-1));
```

This code requires some additional explanation.

- A table of spline coefficients is defined for each dependent variable at the current r with splinefun, a utility in the basic R system.

```
tableS  =splinefun(r,S);
tableIu =splinefun(r,Iu);
tableIh =splinefun(r,Ih);
tableR  =splinefun(r,R);
```

- Three cases for the $\dfrac{dr_u}{dt}$ = drdt are defined.

```
if(ncase==1){drdt=0;}
if(ncase==2){drdt=0.2;}
if(ncase==3){drdt=0.005*S[nr];}
```

For ncase=1, $\dfrac{dr_u}{dt} = 0$ so there is no change in r_u. This case is worth considering since if r_u changes, a programming error is indicated.

For ncase=2, r_u changes at a constant rate 0.2 so that the variation of r_u with t is linear. Again, this case is worth considering since if this response of r_u is not observed, a programming error is indicated.

For ncase=3, drdt is defined by the susceptible population at the outer boundary, S[nr], that is, the velocity equation for the outer boundary.

drdt=0.005*S[nr]

The proportionality constant 0.005 was selected to give an appreciable movement of the outer boundary as the solution evolves in t. Other choices for the moving boundary function can be programmed here. A particular choice will determine how the influenza region outer boundary at $r = r_u$ moves with t.

- The explicit Euler method is used to compute the next r_u.

ru=ru+drdt*dt;

dt is presumed small enough that the Euler method gives sufficient accuracy in the calculation of r_u.

- The grid in r is redefined for the new r_u. In other words, the moving grid is implemented at this point.

r=seq(from=rl,to=ru,by=(ru-rl)/(nr-1));

• Modified solutions S,Iu,Ih,R are computed for the redefined grid r with the previously defined spline coefficient tables, tableS, tableIu, tableIh, tableR. deriv=0 designates the return of the function (rather than its derivatives in r) as interpolated/extrapolated by the spline.

```
#
# Solution on redefined grid
    S  =tableS(r,deriv=0);
   Iu =tableIu(r,deriv=0);
   Ih =tableIh(r,deriv=0);
    R  =tableR(r,deriv=0);
```

This step illustrates the use of an important property of the spline, that is, a different independent variable vector r can be defined and used which permits the implementation of the moving grid. In other words, interpolation/extrapolation with the spline is used to change the solution vectors based on the changed values of r_u and r.

- The current values of r_u and $\dfrac{dr_u}{dt}$ are updated for subsequent plotting.

```
   it=it+1;
   rup[it]=ru;drdtp[it-1]=drdt;
```

`drdtp[it-1]` is used since the vector `drdtp` does not include a value of the derivative at $t = t_0 = 0$ (discussed previously).

- The solutions and t are also placed in arrays for display.

```
#
# Solution for plotting
   for(i in 1:nr){
       Sp[i,it]= S[i];
       Iup[i,it]=Iu[i];
       Ihp[i,it]=Ih[i];
       Rp[i,it]= R[i];
   }
   tp[it]=tout[2];
```

- The next step in t of length dt is initiated (for it<nout).

```
#
# Next step (from while)
   }
```

- The solution is displayed at $t = 0, 300, r = 0, 50, 100$.

```
#
# Display numerical solution
  cat(sprintf(" ncase = %2d, mu = %4.1f\n",ncase,mu));
  for(it in 1:nout){
    if((it-1)*(it-nout)==0){
      cat(sprintf("\n"));
      cat(sprintf("\n        t         r"));
      cat(sprintf("\n                  S(r,t)      Iu(r,t)"));
      cat(sprintf("\n                  Ih(r,t)      R(r,t)"));
    for(i in 1:nr){
      if((i-1)*(i-26)*(i-nr)==0){
      cat(sprintf("\n  %6.0f %6.1f",tout[it],r[i]));
      cat(sprintf("\n   %12.2f %12.2f", S[i,it],Iu[i,it]));
      cat(sprintf("\n   %12.2f %12.2f",Ih[i,it], R[i,it]));
      }
    }
    }
  }
```

- The solutions are plotted in 2D with `matplot`.

```
#
# Plot PDE solutions
#
# 2D
#
# S
  par(mfrow=c(1,1));
  matplot(r,Sp,type="l",xlab="r",ylab="S(r,t)",
    lty=1,main="",lwd=2,col="black");
#
# Iu
  par(mfrow=c(1,1));
  matplot(r,Iup,type="l",xlab="r",ylab="Iu(r,t)",
    lty=1,main="",lwd=2,col="black");
#
# Ih
  par(mfrow=c(1,1));
  matplot(r,Ihp,type="l",xlab="r",ylab="Ih(r,t)",
    lty=1,main="",lwd=2,col="black");
```

```
#
# R
  par(mfrow=c(1,1));
  matplot(r,Rp,type="l",xlab="r",ylab="R(r,t)",
    lty=1,main="",lwd=2,col="black");
```

- The solutions are also plotted in 3D with persp.

```
#
# 3D
#
# S
  persp(r,tp,Sp,theta=120,phi=45,xlab="r",ylab="t",
        zlab="S(r,t)");
#
# Iu
  persp(r,tp,Iup,theta=120,phi=45,xlab="r",ylab="t",
        zlab="Iu(r,t)");
#
# Ih
  persp(r,tp,Ihp,theta=120,phi=45,xlab="r",ylab="t",
        zlab="Ih(r,t)");
#
# R
  persp(r,tp,Rp,theta=60,phi=45,xlab="r",ylab="t",
        zlab="R(r,t)");
```

- The movement of r_u and the associated $\dfrac{dr_u}{dt}$ are plotted against t.

```
#
# Boundary position
  plot(tp,rup,xlab="t",ylab="ru");
   lines(tp,rup,type="l",lwd=2);
  points(tp,rup,pch="o",lwd=2);
  plot(tp[2:nout],drdtp,xlab="t",ylab="dru/dt");
   lines(tp[2:nout],drdtp,type="l",lwd=2);
  points(tp[2:nout],drdtp,pch="o",lwd=2);
```

tp[2:np] is used for t since the derivative array drdtp does not include a value at $t = t_0 = 0$.

- The number of calls to the ODE/MOL routine, pde_1a, is displayed as a measure of the computational effort required to compute the complete solution.

```
#
# Calls to ODE routine
  cat(sprintf("\n\n  ncall = %3d\n",ncall));
```

This completes the discussion of the main program in Listing 4.1. The ODE/MOL routine pde_1a called by lsodes in the main program is the same as in Listing 2.2, and dss004, dss044 are discussed in Appendices A.1 and A.2. Therefore the output from Listings 4.1 and 2.2 is considered next.

4.2   MODEL OUTPUT

The output for ncase=1,2,3 is considered next.

4.2.1   FIXED OUTER BOUNDARY

For mu=0.5, ncase=1, drdt=0, the numerical output appears in Table 4.1.

The solutions from the main programs of Listings 2.1, 3.1 and 4.1 are essentially the same (compare Tables 2.2, 3.3, and 4.1). For example, for comparison, $S(r,t)$ in 2D is in Figs. 2.2-1, 3.2-1, and 4.1-1. This agreement indicates that the MBPDE algorithm (Listing 4.1, ncase=1) gives the same solution for the case of a fixed boundary as the conventional MOL (in Listing 3.1, ncase=3). This special case check is worthwhile since the MBPDE algorithm calls lsodes 16 times while the conventional MOL calls lsodes only once.

The computational effort increases substantially, but it is still manageable (compare with Table 2.2, ncall = 571, and Table 3.3, ncase=3, ncall = 571). The increase in ncall results from 16 starts of lsodes rather than one.

The graphical output is essentially unchanged. For example, compare

- Figures 4.1-1 and 3.2-1 for $S(r,t)$.

- Figures 4.1-2 and 3.2-2 for $I_u(r,t)$.

Table 4.1: Solution of Eqs. (1.1-1)–(1.1-4), (1.2-1)–(1.2-4), and (1.3-1)–(1.3-8), Listing 4.1, ncase=1

```
ncase =  1, mu =   0.5

    t        r
              S(r,t)        Iu(r,t)
              Ih(r,t)        R(r,t)
    0     0.0
                 50.00          5.00
                  0.00          0.00
    0    50.0
                 50.00          0.00
                  0.00          0.00
    0   100.0
                 50.00          0.00
                  0.00          0.00


    t        r
              S(r,t)        Iu(r,t)
              Ih(r,t)        R(r,t)
  300      0.0
                  5.41          0.00
                  0.00         42.77
  300     50.0
                  5.58          0.00
                  0.00         43.74
  300    100.0
                  5.36          0.00
                  0.00         45.69

  ncall = 5279
```

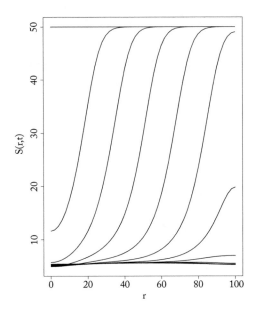

Figure 4.1-1: Numerical solution $S(r, t)$ from Eq. (1.1-1), ncase=1.

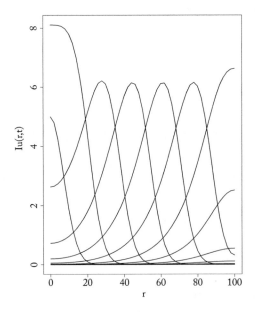

Figure 4.1-2: Numerical solution $I_u(r, t)$ from Eq. (1.1-2), ncase=1.

Figs. 4.1-3 to 4.1-8 are not presented here to conserve space (they are essentially the same as Figs. 3.3-3 to 3.3-8).

The outer boundary position and velocity is indicated in Figs. 4.1-9 and 4.1-10.

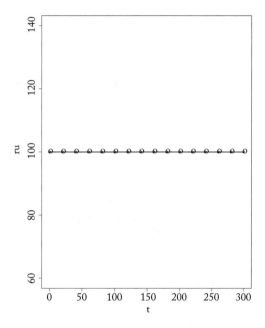

Figure 4.1-9: $r_u(t)$ against t, ncase=1, Listing 4.1.

Figure 4.1-9 confirms $r_u = 100$ for ncase=1.

Figure 4.1-10 confirms $\dfrac{dr_u}{dt} = 0$ for ncase=1.

The preceding output tests the MBPDE algorithm for a fixed boundary. The next case, ncase=2 in Listing 4.1, tests the case of a constant velocity (drdt=0.2).

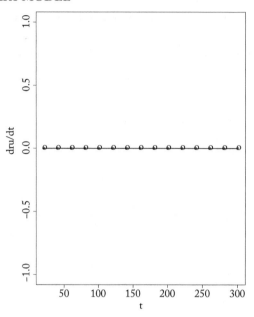

Figure 4.1-10: $\dfrac{dr_u}{dt}$ against t, `ncase=1`, Listing 4.1.

4.2.2   CONSTANT VELOCITY OUTER BOUNDARY

For `mu=0.5`, `ncase=2`, `drdt=0.2`, the numerical output appears in Table 4.2.

The final outer boundary is now at 160 since $drdt = 0.2$ and $r_u = 100 + (0.2)(300) = 160$ (this outer boundary radius appears at $t = 0, 300$ in Table 4.2 since the numerical solution is displayed after it is completed and the vector r has the final 51 values for $0 \le r \le 160$). The boundary movement increased the computational effort to `ncall = 6186`.

The graphical output, Figs. 4.2-1–4.2-10, reflects the effect of the moving boundary (with a single plot for each PDE dependent variable as noted below).

Table 4.2: Solution of Eqs. (1.1-1)–(1.1-4), (1.2-1)–(1.2-4), and (1.3-1)–(1.3-8), Listing 4.1, ncase=2

```
ncase =   2, mu =   0.5

     t         r
               S(r,t)        Iu(r,t)
               Ih(r,t)        R(r,t)
     0     0.0
                   50.00          5.00
                    0.00          0.00
     0    80.0
                   50.00          0.00
                    0.00          0.00
     0   160.0
                   50.00          0.00
                    0.00          0.00


     t         r
               S(r,t)        Iu(r,t)
               Ih(r,t)        R(r,t)
   300     0.0
                    5.41          0.00
                    0.00         42.75
   300    80.0
                    5.59          0.00
                    0.00         44.21
   300   160.0
                    4.91          0.00
                    0.00         47.81


  ncall = 6186
```

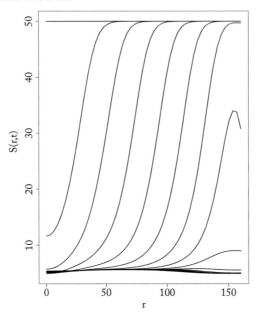

Figure 4.2-1: Numerical solution $S(r, t)$ from Eq. (1.1-1), ncase=2.

A comparison of Figs. 4.2-1 and 4.2-5 for ncase=2 and Figs. 4.1-1 and 4.1-5 for ncase=1 indicates the moving fronts for $S(r, t)$ resulting from the extension of the spatial region from $0 \leq r \leq 100$ to $0 \leq r \leq 160$. Note, however, that a single grid in r is used, $0 \leq r \leq 160$, while the individual solution curves, parametric in t, are for an interval in r that changes with t. For example, for ncase=2,

$$0 \leq t \leq 300; \; dt = (300 - 0)/(16 - 1) = 20$$

$$100 \leq r_u \leq 160; \; dr_u = (160 - 100)/(16 - 1) = 4$$

$$r_u + dr_u = r_u + \frac{dr_u}{dt}dt = r_u + (0.2)(20) = r_u + 4.$$

To indicate the variation in the r grid, individual plots with r as the horizontal (abcissa) variable would be required. For ncase=2, using 16 plots with $dt = 20$, $dr = 4$ the plots would be for

$$t = 0, \; r_u = 100$$

$$t = 20, \; r_u = 100 + 4 = 104$$

$$\vdots$$

$$t = 280, \; r_u = 156$$

$$t = 300, \ r_u = 160.$$

Also, Figs. 4.2-9 and 4.2-10 give a clear indication of the movement of the outer boundary at $r = r_u$.

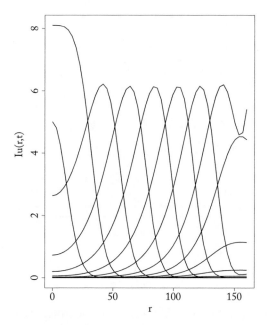

Figure 4.2-2: Numerical solution $I_u(r,t)$ from Eq. (1.1-2), ncase=2.

A comparison of Figs. 4.2-2 and 4.2-6 for ncase=2 and Figs. 4.1-2 and 4.1-6 for ncase=1 indicates the moving pulses for $I_u(r,t)$ resulting from the extension of the spatial region from $0 \le r \le 100$ to $0 \le r \le 160$.

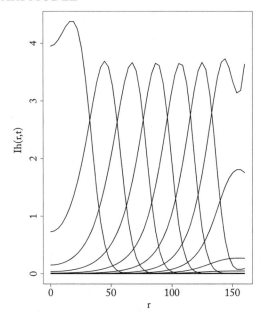

Figure 4.2-3: Numerical solution $I_h(r, t)$ from Eq. (1.1-3), ncase=2.

Figures 4.2-3 and 4.2-7 for ncase=2 indicate the moving pulses for $I_h(r, t)$ resulting from the extension of the spatial region from $0 \leq r \leq 100$ to $0 \leq r \leq 160$.

Figures 4.2-4 and 4.2-8 for ncase=2 indicate the moving fronts for $R(r, t)$ resulting from the extension of the spatial region from $0 \leq r \leq 100$ to $0 \leq r \leq 160$.

Figure 4.2-9 confirms $r_u = 100 + 0.2t$ for ncase=2.

Figure 4.2-10 confirms $\dfrac{dr_u}{dt} = 0.2$ for ncase=2.

The two special cases, ncase=1,2, provide verification of the coding of the moving boundary algorithm programmed in Listing 4.1. The concluding case, ncase=3, is an example of how the outer boundary moves with a velocity that is a function of the PDE dependent variables $(S(r = r_u, t))$.

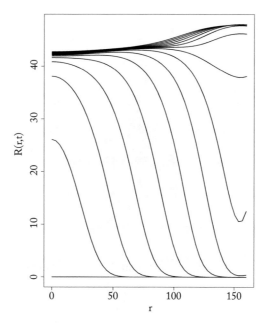

Figure 4.2-4: Numerical solution $R(r, t)$ from Eq. (1.1-4), ncase=2.

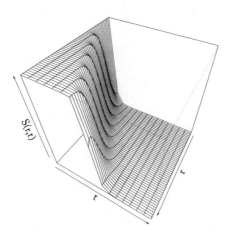

Figure 4.2-5: Numerical solution $S(r, t)$ from Eq. (1.1-1), ncase=2, persp.

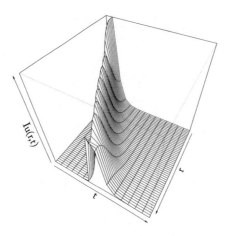

Figure 4.2-6: Numerical solution $I_u(r, t)$ from Eq. (1.1-2), `ncase=2`, `persp`.

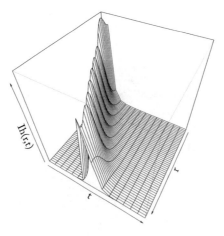

Figure 4.2-7: Numerical solution $I_h(r, t)$ from Eq. (1.1-3), `ncase=2`, `persp`.

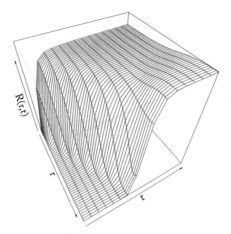

Figure 4.2-8: Numerical solution $R(r, t)$ from Eq. (1.1-4), ncase=2, persp.

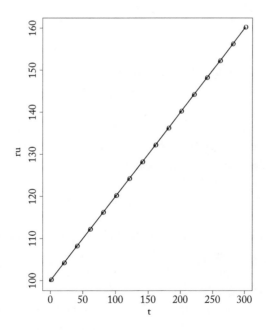

Figure 4.2-9: $r_u(t)$ against t, ncase=2, Listing 4.1.

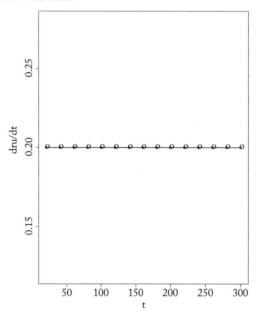

Figure 4.2-10: $\dfrac{dr_u}{dt}$ against t, ncase=2, Listing 4.1.

4.2.3   OUTER BOUNDARY VARIABLE VELOCITY

For mu=0.5, ncase=3, drdt=0.005*S[nr] (Listing 4.1), the numerical output appears in Table 4.3.

The outer boundary is now at 140.1 at $t = 300$ (rather than 160 for ncase=2 so the effect of changing the boundary velocity is clear, and variations in the velocity equation can easily be studied, such as changes in the proportionality constant 0.005 in $drdt = 0.005S[nr]$ for ncase=3). Again, as with ncase =2, at $t = 0$ $r_u = 140.1$ is indicated in Table 4.3 since this numerical output is displayed only after the complete solution is computed (for $t = 0$ to $t = 300$).

The computational effort is ncall = 6137 reflecting the 16 starts (calls) of lsodes.

The graphical output, Figures 4.3-1–4.3-10, reflects the effect of the moving boundary.

Figures 4.3-1 and 4.3-5 for ncase=3 indicate the moving fronts for $S(r, t)$ resulting from the extension of the spatial region from $0 \le r \le 100$ to $0 \le r \le 140.1$.

Figures 4.3-2 and 4.3-6 for ncase=3 indicate the moving pulses for $I_u(r, t)$ resulting from the extension of the spatial region from $0 \le r \le 100$ to $0 \le r \le 140.1$.

Figures 4.3-3 and 4.3-7 for ncase=3 indicate the moving pulses for $I_h(r, t)$ resulting from the extension of the spatial region from $0 \le r \le 100$ to $0 \le r \le 140.1$.

Table 4.3: Solution of Eqs. (1.1-1)–(1.1-4), (1.2-1)–(1.2-4), and (1.3-1)–(1.3-8), Listing 4.1, ncase=3

```
ncase =  3, mu =   0.5

     t        r
              S(r,t)       Iu(r,t)
              Ih(r,t)       R(r,t)
     0    0.0
                  50.00         5.00
                   0.00         0.00
     0   70.0
                  50.00         0.00
                   0.00         0.00
     0  140.1
                  50.00         0.00
                   0.00         0.00


     t        r
              S(r,t)       Iu(r,t)
              Ih(r,t)       R(r,t)
   300    0.0
                   5.41         0.00
                   0.00        42.75
   300   70.0
                   5.62         0.00
                   0.00        43.65
   300  140.1
                   5.10         0.00
                   0.00        46.69
  ncall = 6137
```

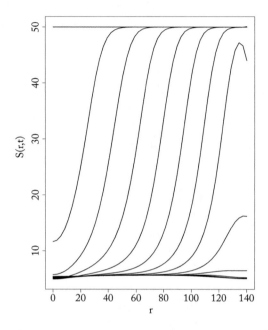

Figure 4.3-1: Numerical solution $S(r,t)$ from Eq. (1.1-1), ncase=3.

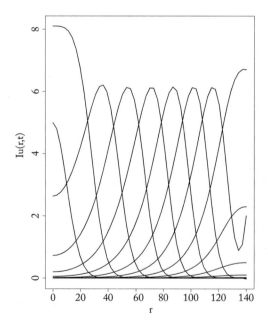

Figure 4.3-2: Numerical solution $I_u(r,t)$ from Eq. (1.1-2), ncase=3.

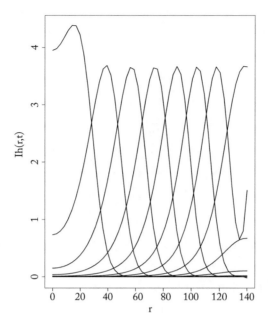

Figure 4.3-3: Numerical solution $I_h(r, t)$ from Eq. (1.1-3), ncase=3.

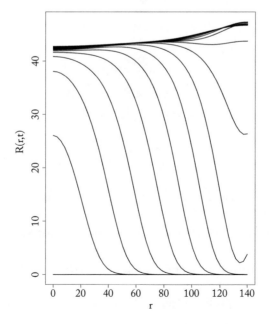

Figure 4.3-4: Numerical solution $R(r, t)$ from Eq. (1.1-4), ncase=3.

Figures 4.3-4 and 4.3-8 for `ncase=3` indicate the moving fronts for $R(r,t)$ resulting from the extension of the spatial region from $0 \leq r \leq 100$ to $0 \leq r \leq 140.1$.

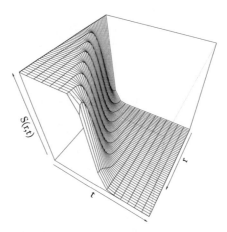

Figure 4.3-5: Numerical solution $S(r,t)$ from Eq. (1.1-1), `ncase=3`, `persp`.

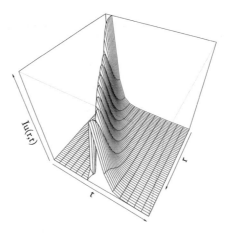

Figure 4.3-6: Numerical solution $I_u(r,t)$ from Eq. (1.1-2), `ncase=3`, `persp`.

Figure 4.3-9 reflects the variable outer boundary from the velocity `drdt=0.005*S[nr]` for `ncase=3`.

Figure 4.3-10 confirms the variable outer boundary velocity `drdt=0.005*S[nr]` for `ncase=3`.

Figures 4.3-9 and 4.3-10 give a clear indication of the movement of the outer boundary at $r = r_u$.

This completes the example of a variable outer boundary of the influenza region.

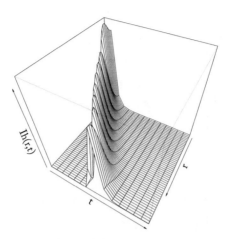

Figure 4.3-7: Numerical solution $I_h(r, t)$ from Eq. (1.1-3), ncase=3, persp.

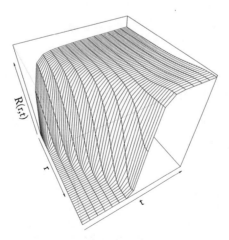

Figure 4.3-8: Numerical solution $R(r, t)$ from Eq. (1.1-4), ncase=3, persp.

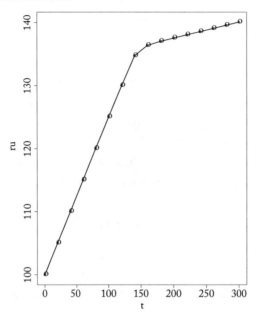

Figure 4.3-9: $r_u(t)$ against t, ncase=3, Listing 4.1.

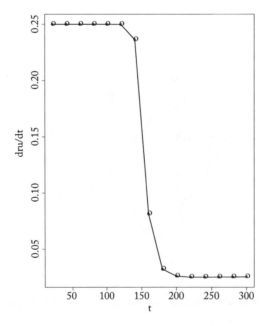

Figure 4.3-10: $\dfrac{dr_u}{dt}$ against t, ncase=3, Listing 4.1.

4.3   SUMMARY AND CONCLUSIONS

The preceding numerical analysis details the PDE model of Eqs. (1.1-1)–(1.1-4), (1.2-1)–(1.2-4), and (1.3-1)–(1.3-8) applied to the case of a moving outer boundary. The MBPDE algorithm is general in the sense that the outer boundary velocity can be defined arbitrarily and therefore can be used to study changes in the size of the influenza region.

The model reported in [2] was selected as a basic introduction to the spatiotemporal modeling of influenza evoluton. The numerical methodology presented in this example can be applied to more general influenza models, and other communicable diseases.

REFERENCES

[1] Soetaert, K., Cash, J., and Mazzia, F. (2012). *Solving Differential Equations in R*, Springer-Verlag, Heidelberg, Germany. DOI: 10.1007/978-3-642-28070-2. 60

[2] Zhang, T. and Wang, W. (2014). Existence of traveling wave solutions for influenza model with treatment, *Journal of Mathematical Analysis and Applications*, 419, pp. 469–495. DOI: 10.1016/j.jmaa.2014.04.068. 85

APPENDIX A

Functions dss004, dss044

A.1 FUNCTION DSS004

Listing A.1: Listing of function dss004

```
  dss004=function(xl,xu,n,u) {
#
# An extensive set of documentation comments detailing
# the derivation of the following fourth order finite
# differences (FDs) is not given here to conserve
# space.  The derivation is detailed in Schiesser,
# W. E., The Numerical Method of Lines Integration
# of Partial Differential Equations, Academic Press,
# San Diego, 1991.
#
# Preallocate arrays
  ux=rep(0,n);
#
# Grid spacing
  dx=(xu-xl)/(n-1);
#
# 1/(12*dx) for subsequent use
  r12dx=1/(12*dx);
#
# ux vector
#
# Boundaries (x=xl,x=xu)
  ux[1]=r12dx*(-25*u[1]+48*u[  2]-36*u[  3]+16*u[  4]-3*u[  5]);
  ux[n]=r12dx*(  25*u[n]-48*u[n-1]+36*u[n-2]-16*u[n-3]+3*u[n-4]);
#
# dx in from boundaries (x=xl+dx,x=xu-dx)
  ux[  2]=r12dx*(-3*u[1]-10*u[  2]+18*u[  3]-6*u[  4]+u[  5]);
  ux[n-1]=r12dx*(  3*u[n]+10*u[n-1]-18*u[n-2]+6*u[n-3]-u[n-4]);
#
```

```
# Interior points (x=xl+2*dx,...,x=xu-2*dx)
  for(i in 3:(n-2))ux[i]=r12dx*(-u[i+2]+8*u[i+1]-8*u[i-1]+u[i-2])
    ;
#
# All points concluded (x=xl,...,x=xu)
  return(c(ux));
}
```

The input arguments are

xl	lower boundary value of x
xu	upper boundary value of x
n	number of points in the grid in x, including the end points
u	dependent variable to be differentiated, an n-vector

The output, ux, is an n-vector of numerical values of the first derivative of u.

The finite difference (FD) approximations are a weighted sum of the dependent variable values. For example, at point i

```
for(i in 3:(n-2))ux[i]=r12dx*(-u[i+2]+8*u[i+1]-8*u[i-1]+u[i-2]);
```

The weighting coefficients are -1, 8, 0, -8, 1 at points i-2, i-1, i, i+1, i+2, respectievly. These weighting coefficients are antisymmetic (opposite in sign) around the center point i because the computed first derivative is of odd order. If the derivative is of even order, the weighting coefficients would be symmetric (same sign) around the center point (as in dss044 that follows).

For i=1. the dependent variable at points i=1,2,3,4,5 is used in the FD approximation for ux[1] to remain within the x domain (fictitious points outside the x domain are not used).

```
ux[1]=r12dx*(-25*u[1]+48*u[2]-36*u[3]+16*u[4]-3*u[5]);
```

Similarly, for i=2, points i=1,2,3,4,5 are used in the FD approximation for ux[2] to remain within the x domain (fictitious points outside the x domain are avoided).

```
ux[2]=r12dx*(-3*u[1]-10*u[2]+18*u[3]-6*u[4]+u[5]);
```

At the right boundary $x = x_u$, points at i=n,n-1,n-2,n-3,n-4 are used for ux[n],ux[n-1] to avoid points outside the x domain.

In all cases, the FD approximations are fourth order correct in x.

A.2   FUNCTION DSS044

Listing A.2: Listing of function dss044

```
dss044=function(xl,xu,n,u,ux,nl,nu) {
#
# The derivation of the finite difference
# approximations for a second derivative are
# in Schiesser, W. E., The Numerical Method
# of Lines Integration of Partial Differential
# Equations, Academic Press, San Diego, 1991.
#
# Preallocate arrays
  uxx=rep(0,n);
#
# Grid spacing
  dx=(xu-xl)/(n-1);
#
# 1/(12*dx**2) for subsequent use
  r12dxs=1/(12*dx^2);
#
# uxx vector
#
# Boundaries (x=xl,x=xu)
  if(nl==1)
    uxx[1]=r12dxs*
          (45*u[   1]-154*u[   2]+214*u[   3]-
          156*u[   4] +61*u[   5] -10*u[   6]);
  if(nu==1)
    uxx[n]=r12dxs*
          (45*u[   n]-154*u[n-1]+214*u[n-2]-
          156*u[n-3] +61*u[n-4] -10*u[n-5]);
  if(nl==2)
    uxx[1]=r12dxs*
          (-415/6*u[   1] +96*u[   2]-36*u[   3]+
            32/3*u[   4]-3/2*u[   5]-50*ux[1]*dx);
  if(nu==2)
    uxx[n]=r12dxs*
          (-415/6*u[   n] +96*u[n-1]-36*u[n-2]+
            32/3*u[n-3]-3/2*u[n-4]+50*ux[n]*dx);
```

```
#
# dx in from boundaries (x=xl+dx,x=xu-dx)
   uxx[   2]=r12dxs*
             (10*u[   1]-15*u[   2]-4*u[   3]+
              14*u[   4]- 6*u[   5]   +u[   6]);
   uxx[n-1]=r12dxs*
             (10*u[   n]-15*u[n-1]-4*u[n-2]+
              14*u[n-3]- 6*u[n-4]   +u[n-5]);
#
# Remaining interior points (x=xl+2*dx,...,
# x=xu-2*dx)
  for(i in 3:(n-2))
    uxx[i]=r12dxs*
            (-u[i-2]+16*u[i-1]-30*u[i]+
          16*u[i+1]   -u[i+2]);
#
# All points concluded (x=xl,...,x=xu)
  return(c(uxx));
}
```

The input arguments are

xl	lower boundary value of x
xu	upper boundary value of x
n	number of points in the grid in x, including the end points
u	dependent variable to be differentiated, an n-vector
ux	first derivative of u with boundary condition (BC) values, an n-vector
nl	type of boundary condition at x=xl 1: Dirichlet BC 2: Neumann BC

nu	type of boundary condition at x=xu
	1: Dirichlet BC
	2: Neumann BC

The output, uxx, is an n-vector of numerical values of the second derivative of u.

The finite difference (FD) approximations are a weighted sum of the dependent variable values. For example, at point i

```
for(i in 3:(n-2))
  uxx[i]=r12dxs*
        (-u[i-2]+16*u[i-1]-30*u[i]+
     16*u[i+1]    -u[i+2]);
```

The weighting coefficients are −1, 16, −30, 16, −1 at points i-2, i-1, i, i+1, i+2, respectievly. These weighting coefficients are symmetic around the center point i because the computed second derivative is of even order. If the derivative is of odd order, the weighting coefficients would be antisymmetric (opposite sign) around the center point.

For nl=2 and/or nu=2 the boundary values of the first derivative are included in the FD approximation for the second derivative, uxx. For example, at x=xl (with nl=2),

```
if(nl==2)
  uxx[1]=r12dxs*
        (-415/6*u[  1]  +96*u[  2]-36*u[  3]+
          32/3*u[  4]-3/2*u[  5]-50*ux[1]*dx);
```

In computing the second derivative at the left boundary, uxx[1], the first derivative at the left boundary is included, that is, ux[1]. In this way, a Neumann BC is accommodated (ux[1] is included in the input argument ux).

For nl=1, only values of the dependent variable (and not the first derivative) are included in the weighted sum.

```
if(nl==1)
  uxx[1]=r12dxs*
        (45*u[  1]-154*u[  2]+214*u[  3]-
        156*u[  4]  +61*u[  5]  -10*u[  6]);
```

The dependent variable at points i=1,2,3,4,5,6 is used in the FD approximation for uxx[1] to remain within the x domain (fictitious points outside the x domain are not used).

Six points are used rather than five (as in the centered approximation for uxx[i]) since the FD applies at the left boundary and is not centered (around i). Six points provide a fourth order FD approximation which is the same order as the FDs at the interior points in x.

Similar considerations apply at the upper boundary value of x with nu=1,2.

Robin boundary conditions can also be accommodated with nl=2, nu=2. In all three cases, Dirichlet, Neumann and Robin, the boundary conditions can be linear and/or nonlinear.

Additonal details concerning dss004, dss044 are available from [1].

REFERENCES

[1] Griffiths, G. W. and Schiesser, W. E. (2012). *Traveling Wave Analysis of Partial Differential Equations*, Elsevier/Academic Press, Boston, MA 92

Author's Biography

WILLIAM E. SCHIESSER

William E. Schiesser is Emeritus McCann Professor of Computational Biomedical Engineering and Chemical and Biomolecular Engineering at Lehigh University.

Index

Printed in the United States
by Baker & Taylor Publisher Services